BLOOD TIES
BEYOND RACE

The unbroken cord that binds us together

Abdisa Kalbesa

Cover design by: Mesfin Bahiru

Interior Design: Barok Solomon

National Library of Australia: 10019761

Printed in Queensland, Australia.

ISBN: 978-0-646-89162-0

www.bloodinfohub.com

NATIONAL LIBRARY OF AUSTRALIA

A catalogue record for this work is available from the National Library of Australia

This book is lovingly dedicated to my partner, Bontu Tolesa, and my aunty, Feben Grosch.

Table of Contents

Table of Contents

Table of Contents

AN IMPORTANT NOTE: This book is intended to provide information and insights based on scientific research and theories surrounding the relationship between blood type and various aspects of human health and identity. However, it is important to note that the content presented in this book should not be considered as medical or genetic advice. Although blood types provide a more precise measure of human variation in comparison to race, it is crucial to exercise caution when utilising them to prevent the reinforcement of preconceptions or discriminatory practises. The utilisation of blood types as a foundation for uneven treatment or stigmatisation is not acceptable.

The characters and events portrayed in this book are mostly fictitious. Any similarity to real persons, living or dead, is coincidental and not intended by the author. The information provided in this book is for educational purposes only and should not be used as a substitute for professional medical advice, diagnosis, or treatment. Readers are encouraged to consult with qualified healthcare professionals or geneticists for personalised advice and guidance regarding their specific health conditions or concerns.

The author and publisher of this book have made reasonable efforts to ensure the accuracy and reliability of the information presented. However, they cannot guarantee the completeness, accuracy, or timeliness of the content. Therefore, the author and publisher disclaim any liability for any loss, damage, or injury caused by the information contained in this book.

By reading this book, readers acknowledge and agree to the above disclaimer and understand that the information provided is not intended to replace professional medical advice or treatment.

Acknowledgement

I am overwhelmed with a profound sense of gratitude as I extend my heartfelt appreciation to the remarkable person who has had an indelible impact on my life - Aunty Feben Dhufera Grosch. Her significance in my journey cannot be overstated, as she has played a pivotal role in the development of the concept I am about to share. Words fail to adequately express the depth of my gratitude for her invaluable contribution, for without her presence and unwavering support, the mere thought of authoring this book would have been unimaginable.

It was during my time studying at a prestigious nursing collage in "Addis Ababa" originally known as Finfinnee, Ethiopia, that I received a precious gift from Feben. She had gone to great lengths to bring a book all the way from Germany, and she graciously bestowed it upon me. This book, titled "Eat Right 4 Your Type," proved to be a transformative turning point in my personal and intellectual journey. Its pages held profound wisdom and insightful guidance on tailoring dietary choices to individual blood types, with the aim of facilitating weight gain or loss. Little did I know that this seemingly innocuous gift would ignite a flame of enlightenment within me, prompting deep contemplation on the potential impact of humanity's connection through blood types.

As I reflect on the journey that led me to this point, I am filled with a profound sense of gratitude for Aunty Feben. Her unwavering support and belief in my potential have been instrumental in shaping the person I have become. Without her guidance and encouragement, I would not have had the courage to embark on this ambitious endeavour. As I embark on this next chapter, I carry her wisdom and guidance with me, knowing that her belief in me will continue to fuel my passion and drive.

As I prepare to share this concept with the world, I am humbled by the magnitude of the task at hand. I am acutely aware of the challenges and criticisms that may arise, but I am also filled with a deep sense of purpose and conviction. I believe that the human diversity based on blood types has the potential to revolutionize the way we perceive ourselves and others. It has the power to bridge divides, foster understanding, and promote holistic well-being.

I am also filled with an overwhelming sense of gratitude as I extend my deepest appreciation to the remarkable individuals of the Gospel Light Christian Church, in Australia. Their unwavering commitment to prayer and their invaluable contributions have played a pivotal role. Their dedication to prayer has created an environment of spiritual strength and guidance, which has been instrumental in my journey. Their prayers have provided me with the strength and resilience to overcome challenges and obstacles that I have encountered along the way. Their unwavering belief in me have been a constant source of inspiration and motivation.

Inside our Church, I have been fortunate to build strong relationships with a significant number of individuals. Their acceptance, love, and support have created a sense of belonging and community that I cherish deeply. The bonds formed within this religious congregation have become a source of strength and encouragement, propelling me forward in my endeavours. I am humbled by the extent of their kindness and generosity, and I am forever grateful for their presence in my life.

In addition to the support of the church community, I must express my profound appreciation to my esteemed spouse, Bontu Tolasa. Her exceptional qualities and unwavering support have been instrumental in my journey. Bontu, 'fedhiikoo,' as I affectionately call her, has been my rock during moments of both happiness and hardship. Her unwavering belief in me and her constant encouragement have been a source of strength and inspiration.

During a time of great difficulty, Bontu's support has been unwavering. Her love and understanding have provided me with the resilience and determination to overcome challenges and continue pursuing my dreams. Her unwavering dedication to our family and her exceptional maternal qualities will leave a profound impact on our future children. They will regard her with admiration and respect, recognising the extraordinary love and care she provides beyond their wildest expectations.

As I reflect on the fortunate circumstances that have shaped my journey, I am overwhelmed with gratitude for the contributions and the unwavering support of my beloved spouse, Bontu. Her presence in my life has been a blessing beyond measure. Her exceptional qualities and unwavering support have not only shaped my personal and professional growth but have also enriched my life in ways that words cannot fully express.

Preface

In the tapestry of human existence, the threads of racial identity have been woven with a heavy hand, casting long shadows over our shared history. Yet, within the intricate weaves lies a universal strand, one that courses through our veins with a silent narrative of its own blood. Not the crimson hue that marks us all as kin, but the distinct types that speak to a deeper connection, one that transcends the superficial divides of skin colour and ancestral lineage. This is the essence of "Blood Ties: Beyond Race," a clarion call to re-examine the foundations upon which we have built our social hierarchies.

Have you ever stopped to consider the life-sustaining fluid that flows within you? It is a marvel of biology, a complex cocktail of cells and plasma that holds the keys to our very survival. Yet, for all its critical functions, it harbors no prejudice, no inkling of racial superiority or inferiority. The heart pumps, the blood circulates, and life goes on, irrespective of the societal constructs that humans have erected.

To grasp the full import of this revelation, we must embark on a journey through time. History, that relentless teacher, has much to impart about the sheer folly of racial divisions. Civilisations have risen and fallen, wars have been waged, and countless lives have been lost, all in the name of racial purity and dominance. But what if this pursuit was misguided from the very start? What if the true measure of our interconnectedness lay not in the colour of our skin but in the shared blood types that bind us all?

Consider the O blood type, the universal donor, a symbol of life and hope in the direst of circumstances. It does not discriminate; its life-giving properties are as potent in the veins of a person from the steppes of Africa as they are in those of an individual from the heart of the Amazon. Where, then, does race factor into this equation of life and shared humanity?

In a similar vein, ponder the rare AB blood type, nature's recipient, accepting transfusions from all others. This blood type, a testament to human diversity, exists in all corners of the globe, further blurring the lines that we have drawn so boldly between races. What does this say about our perceived differences? Are they not, in many ways, an illusion, a construct of the mind rather than a dictate of nature?

The biology of blood types is a fascinating tale of evolution and adaptation. As we delve into the origins of ABO and Rh systems, a mosaic of human migration and genetic inheritance unfolds before us. These markers, which have played a critical role in medicine and forensics, offer a window into our shared past, revealing a story of survival and interconnection that challenges the separatist narrative of race.

Our social fabric, too, is interwoven with tales of blood type. Superstitions and stereotypes have emerged, attributing personality traits and destinies to the bearers of certain blood types. In some cultures, blood type has even become a criterion for matchmaking, employment, and social standing. But is there any scientific merit to these beliefs? Or are they merely another facet of our penchant for categorising and segregating one another?

The significance of blood types cannot be underestimated. But not in the ways that engender division. Rather, they serve as a unifying thread, a reminder that beneath the surface, our similarities outweigh our differences. They compel us to question the validity of race as a construct that has been used to justify discrimination and inequality.

Why, then, do we cling to race as a defining characteristic? Could it be that the narratives of racial difference are so deeply embedded in our psyche that we are blind to the biological truths that unite us? It is a question that demands introspection and honesty.

The journey through "Blood Ties: Beyond Race" is not merely an academic exercise; it is a voyage towards a greater understanding of what it means to be human. It is an invitation to dismantle the barriers we have built and to see one another not as disparate races but as fellow travellers on the path of life, connected by the very blood that gives us life.

In Book, we delve into the fascinating world of blood types and their impact on various aspects of our lives. While much attention has been given to the role of race in shaping human characteristics, this book takes a unique perspective by exploring the lesser-known influence of blood types.

Drawing on extensive research and scientific studies, this book uncovers the intricate connections between blood types and their effects on our health, personality traits, relationships and even our dietary needs. We go beyond the commonly known ABO blood types and delve into the complexities of the Rh factor, exploring how these blood type variations can shape our lives in unexpected ways.

Through engaging storytelling and accessible explanations, this book reveals the distinct characteristics and tendencies associated with each blood type. From the meticulous and cooperative nature of Type A individuals to the creative and adaptable traits of Type B individuals, readers will gain a deeper understanding of themselves and others around them.

Moreover, this book explores the implications of blood types in various aspects of life, including relationships, dietary choices, and even personal development. By recognising and embracing our unique blood type characteristics, we can navigate life's challenges more effectively and foster better understanding and harmony in our interactions with others.

"Blood Ties: Beyond Race" also delves into the role of blood types in determining susceptibility to certain diseases and conditions. We explore the latest scientific findings on how blood types can influence our immune system, digestive health, and overall well-being. Armed with this knowledge, readers can make informed decisions about their lifestyle choices, including diet, exercise, and preventive healthcare measures.

Throughout the book, we emphasise the importance of embracing diversity and dispel any notions of superiority or inferiority based on blood types. Instead, we celebrate the rich tapestry of human diversity and encourage readers to appreciate the unique qualities that each blood type brings to the table.

In a world where understanding and acceptance are more crucial than ever, this book offers a fresh perspective on human diversity. By shedding light on the hidden influences of blood types, this book empowers readers to embrace their individuality, foster empathy, and build a more inclusive society.

This book is more than a collection of facts and figures; it is a manifesto for a new era of human relations, one in which blood types, not race, become a symbol of our common humanity. It is a call to action, to embrace the scientific truth and to let go of the antiquated notions that have divided us for far too long.

Are your eyebrows raising in scepticism? Does the thought of blood types carrying more weight than race seem too avant-garde, or perhaps, too simplistic? I recognise the complexity of the social fabric that has been meticulously woven over centuries. But as a medical science student, I've learned to look beyond appearances, to seek the truth that lies at the cellular level. Let me allay your doubts by revealing the compelling science and enthralling stories that have been obscured by the shadows of ignorance and fear. Trust in the evidence, the anecdotes, and the undeniable logic that will be meticulously laid before you.

Now, picture yourself not just as a reader, but as a tourneyer. You are about to navigate the rivers of history, culture, and ethics, each turn revealing a new facet of our shared humanity. Feel the pulse of the collective heartbeat, the rhythm that binds you to the person next to you, regardless of the colour of your skin. Can you imagine the transformative power of such a perspective? Can you sense the crumbling of barriers, the dissolving of misunderstandings, as you come to see your neighbour, your friend, or even a stranger as kin?

This voyage will not be one of passive observation. I am asking for your engagement, for your mind and heart to be as open as the veins that will lead us to enlightenment. By the end of this book, you will not only recognise the value of our shared blood but will be imbued with a newfound sense of connection to all of humankind. This is my commitment to you: to present a narrative so compelling, a vision so clear, that your perception of race and kinship will be forever altered.

Let us begin with a promise: by the time you reach the final page, you will have gained not just knowledge, but wisdom—the wisdom to see beyond the superficial, to understand the true nature of our bonds. This book is an invitation to join me, on a ground-breaking journey. Together, we will explore the powerful and often overlooked ties that bind us. Are you ready to discover the deep connections that race cannot define, that geography cannot confine, and that, until now, history has not dared to align?

With every heartbeat, we affirm our existence, our resilience, and our shared destiny. The symphony of survival beckons, and its melody resonates with the truth of our blood ties. Beyond the construct of race lies a profound unity, a kinship that courses through our veins. This is the journey we embark upon—a journey into the heart of what it means to be human.

As we turn the pages, let us do so with open minds and hearts, ready to challenge the status quo and to forge a future where blood type matters beyond the realm of racial divides, define our shared destiny. For in the end, it is not the colour of our skin that matters, but the content of our character and the blood that flows within us all.

CHAPTER

1

What is blood?

The vital fluid that circulates inside the human body is referred to as blood. The assessment of an individual's health status necessitates the use of blood samples. Within the field of medicine, the process of blood transfusion holds significant importance. Medical practitioners usually study this type of tissue more frequently than any other tissues in order to find the root cause of a patient's ailment.

Symbolic representation of life is often depicted through the presence of blood. The presence of an intelligent creator and master craftsman may be inferred from the characteristics of blood. The study of blood allows us to explore the significance behind our existence. The viability of the human organ is contingent upon several constituents, among which blood is a crucial element.

Blood, in conjunction with other constituents of the cardio-vascular system, including blood vessels and the heart, is regarded as one of the most crucial elements. The circulatory system plays a crucial role in the transportation of oxygen and nutrients to every cell in the human body. Additionally, it facilitates the removal of waste products, including carbon dioxide, from the cells for excretion.

The notable scientists, William Harvey and Anton van Leeuwenhoek, made substantial advancements in our comprehension of cardiac function and blood circulation, while also unveiling novel insights into several facets of haematology. The origins of these findings may be attributed to the period during which these two individuals were living. Blood serves as the fundamental sustenance for life, exerting a significant influence on an individual's lifespan. According to Harvey, blood is the primary entity to initiate life and the ultimate element to cease existence, it's a notion with which I concur.

Blood: The River of Life

In the grand theatre of the human body, blood is the tireless performer, the unsung hero in the drama of existence. It surges through our veins and arteries with the quiet strength of a river carving its path through the earth, nurturing and sustaining life at every turn. Every drop is a microscopic ecosystem, teeming with cells, proteins, and platelets—a cocktail of life that speaks to the complexity and wonder of the human form.

Let us begin by drawing back the curtain on this river of life, revealing the intricacies of its composition. Blood is not merely a red liquid, but a dynamic and multifaceted substance composed of plasma, red and white blood cells, and platelets. Plasma, the golden-hued liquid, accounts for about 55% of blood's volume and serves as the medium through which cells and other substances travel. Suspended within it are the red cells, the carriers of oxygen, giving blood its distinctive colour, white cells, the guardians of our immune system, and platelets, the architects of healing, orchestrating clot formation.

The concept of blood as a life-sustaining fluid is familiar, but the depth of its role is profound. It is a transporter, a regulator, a defender, and a healer. Blood ferries essential nutrients and oxygen to the remotest outposts of the body. It regulates temperature and pH balance, keeping the internal environment in a state of equilibrium. It defends against invading pathogens, with white cells acting as sentinels, ever vigilant. And when the skin is breached, it is blood that answers the call, platelets assembling to mend the tear and restore integrity.

To illustrate this, consider the case of Qananiisa aka Kenenesa, an Ethiopian marathon runner, his muscles screaming for oxygen as he pushes towards the finish line. Here, blood demonstrates its role as a supplier, delivering oxygen and fuel while whisking away the waste products of exertion. Or take the example of a child with a scraped knee, where blood's healing properties spring into action, coagulating to form a protective scab. These scenarios are but two threads in the rich tapestry of blood's functions.

From an anthropological perspective, blood has been venerated and feared, symbolising both life and death, purity, and pollution. In many cultures, it represents vitality and strength, and sharing blood is seen as a bond stronger than any written contract. In the medical realm, blood transfusions are a testament to our interconnectedness, a process where the essence of one individual can literally revive another.

But what do the numbers say? Consider that every day, the human heart pumps approximately 2,000 gallons of blood, circulating the body's entire supply around 1,440 times. These statistics are not mere trivia; they underscore the relentless work blood performs in service to our survival.

Diving deeper, we unravel the jargon of hematology. Haemoglobin, the protein in red cells that binds oxygen, is more than just a biological term; it is the key to our vitality. Leukocytes, the white cells, might seem like foreign agents, but they are the skilled warriors in the battle against infection. Understanding these terms is not just educational—it is an act of empowerment, a step towards demystifying our own bodies.

As we draw this exploration to a close, let us distil the essence of our findings. Blood is the river of life, an entity that embodies the paradox of being universal and unique simultaneously. It is a conduit of life, but also a canvas where the story of our health, heritage, and humanity is painted.

Have you ever stood in awe of the relentless journey each cell undertakes, a microscopic odyssey that is both ordinary and miraculous? Have you considered the silent symphony played out within you at every moment, a testament to the marvel that is the human body?

In pondering these questions, remember that blood, in all its complexity and simplicity, is a narrative written in the language of life. As this topic of our exploration concludes, remember that the story of blood is the story of us—written not with ink, but with the very substance that sustains us. It is a tale that continues to unfold with each heartbeat, a story of life, of struggle, and of the enduring connections that bind us all.

The Spiritual Significance of Blood

Blood pulses at the very heart of existence, a crimson thread weaving through the tapestry of life. It is a universal emblem, one that resonates deeply within the human psyche, carrying with it a cargo of profound meanings and associations. But what is it about this vital fluid that elevates it to the realm of the spiritual?

In its essence, blood is the life-sustaining substance that courses through the veins of living creatures, a complex elixir composed of cells, nutrients, and proteins. Yet, this clinical description scarcely scratches the surface of blood's symbolic potency.

The significance of blood extends far beyond its biological function. It is emblematic of life and vitality, often regarded as the seat of the soul or the locus of life force itself. Throughout history, blood has held a central place in the rituals and beliefs of countless cultures, symbolising concepts as diverse as kinship, sacrifice, and redemption.

Delving into the annals of time, one might uncover the etymological roots of the word 'blood' in various languages, each echoing a similar reverence. The Old English 'blōd' and the German 'Blut' speak to ancestral connections, hinting at blood's role in defining relationships and lineage.

To contextualise blood within a broader spiritual framework is to acknowledge its ubiquity in religious and mystical traditions. In Christianity, the blood of Christ signifies the ultimate sacrifice and the promise of salvation. The wine in the Eucharist transforms into a symbol of this sacred blood, a ritual echoed in various forms across the spiritual spectrum.

In the beliefs of indigenous Africans, for instance, the Oromo people of Ethiopia, blood is often an offering to the spirits or deities, a means to appease or communicate with forces beyond the visible world. Though not all individuals, who identify as Oromo, hold this belief. The ancient Mayans, for instance, practiced bloodletting as a way to nourish the gods and ensure the continuity of the cosmos.

The applications of blood as a spiritual symbol are as varied as they are profound. In Vodou, it is an essential component in connecting with the Loa, the spirits of Haitian religion. In Hinduism, blood sacrifices, though less common today, were historically part of certain rituals to honour the gods and seek their favour.

One might question, then, whether this potent symbol carries misconceptions within its crimson depths. It is crucial to distinguish between the symbolic and the literal, for not all references to blood in spiritual contexts advocate for physical acts. The metaphorical use of blood often transcends the material, pointing towards a deeper truth or principle.

In the modern world, the language of blood continues to permeate our understanding of connection and identity. Phrases like 'blood is thicker than water' encapsulate the enduring belief in the bond of family and the precedence it holds over other relationships.

But, what of the blood that flows not through veins but through the veins of society? The metaphoric blood spilled in struggles for justice, the bloodlines that chart our ancestry, or the 'blood' that fuels our passions and endeavours? Each is a testament to the enduring symbolism of blood, as it courses beyond the literal and into the figurative heart of our existence.

Consider, for a moment, the blood that symbolises the shared humanity that courses through each of us. Does it not remind us of our collective vulnerability, our innate need for connection and understanding? Does it not prompt the question of what it truly means to be kin, beyond the limitations of race or creed?

In a world often divided by superficial differences, the symbol of blood beckons us to look deeper, to recognise the common thread that binds us all. It is a call to transcend the physical and embrace the spiritual kinship that unites us beyond race, beyond borders, beyond the very concept of 'otherness.'

Blood, therefore, is not merely a symbol within the spiritual realm; it is an invitation. An invitation to explore the interconnectedness of life, to acknowledge the sacred within the mundane, and to embrace the profound truth that, in the grand tapestry of existence, we are all woven together by threads as enduring and as vital as the blood that gives us life.

Sacred Texts and the Sanctity of Blood

The essence of blood has long captivated the human imagination, its significance reaching into the deepest spiritual recesses of our minds. This life-giving fluid, traversing the delicate channels of our bodies, has also found a sacred place within the pages of hallowed texts, where its symbolism has been eternally inscribed.

In the beginning, these sacred scriptures depict blood as more than a mere constituent of flesh. It is portrayed as the divine conduit through which the breath of life flows. From the Judaic covenant marked by circumcision to the Islamic rituals of sacrifice during Eid al-Adha, blood surfaces as a testament to faith and a marker of devoutness.

Journey with me as we delve into the ancient scrolls and explore how blood has been revered, sanctified, and even feared. The Bible, the Torah, the Quran, and the Bhagavad Gita—each of these venerable compendiums of wisdom offers unique insights into the spiritual significance of blood.

Consider, for instance, the narrative of Passover in Judaism: the lamb's blood smeared on doorposts as a signal for divine mercy and protection. Here, blood serves as a shield, a silent guardian that speaks of deliverance and survival. Similarly, in Christianity, the crucifixion of Jesus Christ and the shedding of His blood become central to the theme of redemption.

This act transforms the understanding of blood from a life-source to a symbol of eternal life, a currency for the salvation of souls.

The symbolism of blood in Christianity is of great importance, since it signifies several facets of the faith, such as sacrifice, redemption, and the covenant between God and humanity. The biblical narrative holds a profound connection to the notion of blood, which is extensively examined in terms of its spiritual implications through various rites, teachings, and theological interpretations.

The notion of blood is intricately linked to the act of sacrifice. The symbolism discussed in this context finds its roots in the Old Testament, when the act of offering animal sacrifices was employed as a method of seeking forgiveness for sins. The act of offering the blood of sacrificial animals, such as lambs, was seen as an essential ritual to achieve purification and forgiveness from the people's transgression.

The concept of sacrifice imagery reaches its pinnacle in the figure of Jesus Christ, commonly known as the "Lamb of God." Jesus was voluntarily presented himself as the flawless offering to atone for the transgressions of mankind. The shedding of blood on the cross is often seen as the paramount act of atonement, serving to offer salvation and foster peace between God and the human race.

The sacrament of the Eucharist, sometimes referred to as Holy Communion or the Lord's Supper, serves to further underscore the significance of blood symbolism. During this religious ceremony, adherents of Christianity engage in the consumption of bread and wine, which are said to undergo a transformation into the physical embodiment of Christ's flesh and blood.

The wine, which symbolises the blood of Christ, is often seen as a potent emblem of the new covenant established between God and humanity. Similar to the act of Jesus sharing the cup of wine with his disciples, Christians engage in the sacrament of the Eucharist as a means of commemorating and actively engaging in the sacrificial nature of Christ.

The correlation between the shedding of blood in Christianity and the notions of redemption and forgiveness is highly intertwined. The prevailing notion posits that by means of Jesus' sacrifice and the shedding of His blood, humanity is presented with the prospect of redemption from sin and the restoration of harmonious relations with God.

According to the biblical account, in the Old Testament, God instructed the Israelites to abstain from consuming blood, as it is considered to be the essence of life and should not be consumed in conjunction with flesh. According to Deuteronomy 12:23–24, there exists a prohibition against consuming blood, instead requiring it to be poured out upon the ground in a manner similar to water.

According to the scriptural passage found in Leviticus 17:10–12, those who consume blood, whether they are Israelites or foreigners dwelling among them, shall face divine retribution in the form of being ostracised from the community. This passage highlights that the divinely appointed manner of atonement required the eating of blood.

These examples, poignant as they are, merely scratch the surface of our exploration. Blood's depiction in sacred texts often intertwines with the rituals and observances fundamental to spiritual practice. The Hindu tradition, for instance, tells of the goddess Kali, often depicted with a garland of severed heads and a skirt of arms, her tongue dripping with blood. In this imagery, blood symbolises the destruction of ego and ignorance, a metaphor for the transformative power of the divine.

And yet, the sanctity of blood is not without its complexities. While some texts exalt its spiritual significance, others caution against the misuse of this potent symbol. The Quran, for example, emphasises that the sanctity of human life supersedes the blood spilled in sacrifice, urging followers to embrace the spiritual intent rather than the physical act.

These divergent perspectives invite us to ponder the multifaceted nature of blood within a religious context. They raise questions about the nature of sacrifice and the value of life. How does one reconcile the reverence for blood with the injunctions against its shedding? Where is the line drawn between sacred symbolism and literal interpretation?

In seeking answers, we must also consider the empirical evidence that has emerged over the centuries. Historical accounts and archaeological findings have shed light on how these spiritual traditions have evolved, influencing, and being influenced by the understanding of blood. The data reveals a tapestry of beliefs and practices, each thread coloured by the cultural and temporal contexts in which they were woven.

To unravel the complexity of terms like 'sacrifice' and 'covenant' is to engage in a linguistic excavation, peeling back the layers of meaning accrued over millennia. The original

Hebrew word 'korban,' often translated as 'sacrifice,' encompasses a broader notion of drawing near to the divine. This etymological insight offers a reframed understanding of blood's role—not as a commodity for barter with Gods, but as a medium for spiritual communion.

As we draw this examination to a close, let us reflect on the key insights gleaned from our journey through sacred texts. Blood, in its rich symbolism, represents the paradoxes of life—creation and destruction, mortality and eternal life, the physical and the metaphysical. It challenges us to look beyond the visceral and to perceive the spiritual threads that bind us in a shared human experience.

In the end, the sanctity of blood in sacred texts serves as a mirror, reflecting our own search for meaning and connection. It is a reminder of the profound truth that, despite our diverse paths, we are all part of a larger, transcendent narrative—one in which every drop of blood tells a story of life, of sacrifice, and of the unending quest for the divine.

Ritualistic Bloodletting: A Historical Examination

Blood's journey through the annals of history is as riveting as it is macabre. Deep within the caverns of time, this crimson stream has carved its way into the bedrock of civilisations, emerging as a potent symbol of both life and death, community and the divine. Its essence, so intrinsically linked to our own, pulses through the collective heartbeat of humanity's past.

In the earliest dawns of human society, bloodletting was not merely an act of faith but a conduit to the supernatural. What stirred in the hearts of our ancestors as they first made incisions into flesh to release the sacred fluid? Imagine the primal scene: under the wide canopy of a starlit sky, the shaman's blade glistens, a sliver of moonlight caught on its edge. The air is thick with anticipation as the tribe gathers, their eyes reflecting the flickering flames of a ceremonial fire.

The genesis of these rites remains shrouded in mystery, the specifics lost to the unforgiving erosion of time. Yet, artifacts and cave paintings whisper of these early practices. In regions as diverse as Mesopotamia, the Americas, and Sub-Saharan Africa, evidence points toward a shared belief in the metaphysical properties of blood.

As we traverse the timeline, etchings on ancient tombs and temples reveal more. The Maya, a civilisation renowned for their astronomical prowess, partook in bloodletting to communicate with their gods and ancestors. One cannot help but marvel at the intricate glyphs depicting kings and queens piercing their tongues, earlobes, or even genitals to offer their blood to the deities. The pain, one imagines, was an integral part of the ritual, a physical sacrifice to ensure cosmic harmony.

In what ways did these ceremonials diverge across the oceans? The pictograms of Egypt speak of a subtler bloodletting ritual, one that invoked the protection of the gods without the dramatic spillage seen in Mayan culture. It was a nuanced dance with the divine, where the intent overshadowed the volume of blood offered.

Fast forward to the Greco-Roman world, where bloodletting takes on a more clinical guise. The practice, now infused with the emerging principles of medicine, becomes a common treatment for various ailments. Hippocrates, the father of medicine, and later Galen would argue for the balancing of humours, wherein bloodletting was believed to restore bodily equilibrium. Did they perceive a link, however tenuous, to the spiritual bloodletting of their forebears?

Contemplate for a moment the physician's lancet as it pierces the patient's vein. Here, the ritual is stripped of communal spectacle, yet remains deeply intimate. The act is not for the gods but for the individual's well-being. Still, does not the hope for healing echo the ancient desire for divine favour?

The Renaissance would see a resurgence of bloodletting, albeit steeped in scholarly debate and anatomical discovery. No longer the realm of mystics, it had become a subject of scientific inquiry. How curious it is that as our understanding of the human body deepened, so too did our reliance on this age-old practice, a testament to its enduring allure.

The imagery is stark and often unsettling: the flam, the scarificator, the leeches. These are the tools that have drawn out life's essence in the name of health. Yet, even as the practice reached its zenith in the 18th and 19th centuries, the winds of change were stirring. The age of enlightenment would soon cast a critical light on bloodletting, shifting the tides towards modern medicine.

How, then, has bloodletting evolved in our contemporary world? It may surprise some to learn that the practice, though no longer mainstream, persists in various forms. Therapeutic phlebotomy, for example, is a medically endorsed procedure to treat certain conditions, such as hemochromatosis. Here, the removal of blood is not ritualistic but rooted in empirical necessity. And yet, can we wholly divorce it from its spiritual antecedents?

Perhaps more direct descendants of ritualistic bloodletting are found in the practices of some modern-day indigenous tribes or in the rites of certain religious ascetics. Here, blood is once again a token of faith, a physical expression of a spiritual quest. It is both a nod to tradition and an adaptation to the present.

Controversy has never been far from bloodletting, its history marked by moments of fierce opposition and radical support. One can't help but ponder the ethical dimensions: where should we draw the line between cultural heritage and the sanctity of the body? The debate is complex, riddled with shades of grey that defy simple resolution.

As we draw the curtains on this exploration, a lingering question remains: what is it about blood that continues to hold such sway over our imaginations? This vital fluid, coursing through the annals of history, bears witness to our collective narrative—a tale of evolving beliefs, practices, and knowledge.

In reflecting on the journey of bloodletting, from the spiritual to the scientific and back again, we are reminded of the cyclical nature of human experience. It is a narrative not of linear progress but of ebb and flow, a rhythm that echoes the very pulse of life itself. For in every drop of blood, there lies a story, a piece of our shared humanity, ever unfolding in the vast tapestry of time.

Rituals of Blood: A Global Perspective

In the vast expanse of human history, our fascination with blood has painted a crimson thread, weaving through the fabric of our cultural tapestries. From the lofty peaks of the Andes to the dense forests of West Africa, blood has served as a cornerstone for rituals and rites, each culture imbuing it with a unique significance that begs for exploration. The purpose of this comparative study is not to judge or to horrify, but to understand—to peer into the shared well of humanity and come away with a deeper comprehension of what unites us, and what sets us apart.

We establish our criteria for comparison at the intersection of myth and ritual. How do cultures use blood in their sacred ceremonies, and what does this usage say about their relationship with the divine, the natural world, and the fabric of their society? It is here, in these questions, that we begin our analysis.

In the highlands of Nepal, we find the indigenous practice of buffalo sacrifice during the festival of Dashain, a direct appeal for blessings from the goddess Durga. Here, blood is a symbol of fertility and thanks, an offering to ensure the continuity of life and the favour of the gods. Contrast this with the Aztec rituals of human sacrifice, where blood was the currency of sustenance for the sun god, a necessary tribute to ensure the sun's rising and the world's continuity.

Delving into these practices, one cannot help but marvel at the stark differences and the underlying similarities. Both cultures saw blood as a means of continuation, yet their offerings differed vastly—one animal, the other human. The question arises: does the nature of the sacrifice reflect how a culture views life itself?

Visual aids, such as the intricate carvings that adorn the temples of Kathmandu, illustrating the buffalo sacrifice, or the stone reliefs of Tenochtitlan depicting the heart-extraction ceremony, serve to bring these comparisons into sharper focus. Each image captures a moment of connection between the mortal and the divine, an attempt to bridge the gap with the most sacred fluid of all.

These rituals, shocking as they may be to the uninitiated, are not relics of a barbaric past, but rather, they provide insights into the broader implications of blood as a symbol. The buffalo's lifeblood nourishes the soil, just as the blood of the sacrificed was believed to feed the gods and maintain the cosmic balance.

Our contemporary relevance is found when we consider how these rituals continue to influence modern practices. The Oromo people of Ethiopia, for instance, have a unique belief and practice when it comes to resolving disagreements and restoring peace. According to their tradition, if two individuals had a disagreement that cannot be resolved through other means, they will engage in shading an animal's blood to make peace.

In this ritual, the individuals involved in the dispute come together and intentionally shed blood as a symbolic act of reconciliation. The act of spiling an animal blood is seen as a powerful gesture that signifies a willingness to let go of grievances and restore harmony.

This belief in the power of blood to restore peace reflects the deep cultural values and traditions of the Oromo people. It highlights their emphasis on communal harmony, forgiveness, and the importance of resolving conflicts in a peaceful manner.

In Nepal, animal sacrifice has become a point of contention, a clash between tradition and a growing animal rights movement. In Mexico, the legacy of the Aztecs is evident in the cultural pride and the lingering fascination with death and rebirth, seen during the annual Día de los Muertos celebrations.

But let us turn our gaze now to a less discussed, yet equally profound, blood ritual—the scarification ceremonies of the Sepik River tribes in Papua New Guinea. Here, young men are initiated into manhood through an arduous process of skin cutting, designed to replicate the tough, scaled skin of a crocodile. Blood is shed not as an offering to deities but as a rite of passage, a physical transformation that marks a psychological metamorphosis.

How does this compare to the bloodletting rituals of the Mayans, who saw the act as a conduit for communication with the gods, a way to seek visions and divine favour? Both use the body as a canvas, yet the intentions diverge—one seeks transformation, the other, revelation.

Ponder this: does the act of spilling one's own blood for a higher purpose not speak to a universal search for meaning, for a connection to something greater than oneself?

As you absorb these stories, allow yourself to be transported across oceans and epochs, to lands were blood flows with sacred intent. And in the quiet moments, ask yourself—how do the rituals of your own life reflect the ancient pulse that beats within us all?

The journey is far from over, and as we continue, remember that each drop of blood spilled in the name of tradition is a testament to our shared humanity. For it is through these rituals that we touch the heart of what it means to be human, bound by blood, beyond race, across the endless expanse of time.

Blood and the Essence of Being

Within the intricate fabric of human understanding, the concept of blood as the vessel of the soul or essence has been interwoven into our collective psyche, transcending the physical realm and entering the philosophical. It is here that our exploration continues, delving into the profound connections between blood and the essence of being, examining this relationship through a lens that seeks to distil truth from tradition, science from superstition.

Blood, in its most primal sense, is the life-sustaining river that flows through the valleys and peaks of our bodies, a crimson current charged with the responsibility of nourishment and vitality. But beyond this corporeal function, lies a deeper, more enigmatic significance. Across cultures and epochs, blood has been revered as the harbinger of the soul, an ethereal substance that carries the very core of who we are.

At the heart of this discourse is the assertion that blood is more than a mere biological necessity; it is the conduit through which our truest selves are expressed and understood. The evidence for this claim is as diverse as it is compelling, encompassing a spectrum of disciplines from the spiritual to the empirical.

Consider the poignant words of the philosopher and poet, Rumi: "The soul is here for its own joy." Such a statement echoes the sentiment that the soul's journey is paramount to our existence. How then does blood play its part in this odyssey of joy? Scholarly works suggest that in many traditions, the soul is believed to reside in the blood, a belief that has shaped rituals and rites across millennia.

The ancient Greeks, for instance, viewed blood as the seat of life and, by extension, the repository of the psyche. Aristotle posited that the heart, brimming with blood, was the centre of sensation and the soul's abode. This idea found resonance in the practice of blood sacrifices, where the spilling of an animal's blood was seen as an offering of its life force to the gods.

But let us delve deeper into this evidence and its implications. The Vedic texts of India speak of blood as the 'sap of life,' a powerful force inextricably linked to the divine essence that animates all beings. In these ancient scriptures, blood is both literal and metaphorical, a symbol of the life energy that connects the microcosm of the individual to the macrocosm of the universe.

Venturing into the realm of counterevidence, one must acknowledge the perspectives of modern science, which positions blood as a complex tissue comprised of cells and plasma, its functions confined to the physical. The soul is not within the purview of science's reach, a matter relegated to the philosophical or theological.

Yet, in the rebuttal to such empirical scepticism, one could argue that the essence of being is not something that can be distilled through a microscope or measured with a syringe. The intangible nature of the soul demands a different kind of inquiry, one that accepts the limitations of scientific scrutiny when it comes to the metaphysical.

Further supporting evidence can be found in the narratives of those who have experienced profound moments of connection or transformation through blood-related experiences. Tales of life-saving transfusions, bloodline revelations, and even near-death episodes often reverberate with a newfound understanding of existence and self.

As we draw the threads of this discussion tighter, the reinforced assertion is clear: blood, in its myriad representations, remains a potent symbol of the essence of being. It is a bridge between the tangible and the intangible, the known and the unknowable, the self and the soul. Our quest to comprehend its full significance may be as boundless as the soul itself, but it is a journey that promises insight into the most profound aspects of our existence – a quest for understanding that, like the coursing of blood through our veins, is ever ongoing and infinitely vital.

CHAPTER

2

Cultural Currents of Blood

In the tapestry of human history, threads of bloodlines weave a complex pattern that transcends the mere concept of race. The chronicles of ancestry sing of bonds that surpass the superficial, delving into the essence of kinship. Let us journey to the heart of ancient civilisations, where the notion of blood as a marker of identity first emerged.

Travel back in time to the empires of old, where lineage determined one's destiny. In the sprawling courts of Egyptian pharaohs, blood was revered as divine, a celestial river channelling the power of Gods through the veins of royals. The Spartans, valiant in their militarism, saw blood as the currency of honour, a testament to their unyielding spirit in battle. Across the oceans, Mayan priests offered their own lifeblood to the gods, believing it to be the most sacred of sacrifices.

These historical milestones, where blood was both a literal and symbolic life force, have directly influenced our modern understanding of inheritance and identity. The blood-feuds of medieval clans, the bloodlines that shaped the aristocracy of Europe, and the blood quantum laws that would later categorise the indigenous peoples of Australia all stem from this ancient obsession with blood as the carrier of legacy.

From past to present, the implications of these practices resonate in today's discourse on heritage and race. Can we trace our own lineage with the same certainty as a genealogist tracing royal descent through centuries-old parchments?

Why does history matter now, in this era of genetic testing and multicultural societies? The answers lie in the blood-soaked soil of battlefields where ancestors fought, in the ink of treaties that redrew maps, and in the shared blood of people from all corners of the world. Understanding our blood ties is to understand the interconnectedness of humanity, to recognise that our differences are merely ripples on the surface of a deeper, shared existence.

As we segue to the story of our times, where DNA tests promise insights into our ancestral pasts, we must ponder the implications. Do these revelations bring us closer together, or do they reinforce old divisions? The blood that once determined the fates of empires now holds the keys to personal histories, waiting to be unlocked.

Delve deeper into the narrative. Amidst the ancient ruins, can you hear the whispers of your ancestors? Do you feel the weight of their legacies in your bones, or does the wind carry away their stories, untold and forgotten?

Imagine standing upon the battlegrounds of yore. Do the fallen warriors speak to you of valour, or do they mourn the bloodshed that history often glorifies?

It is said that those who do not learn from history are doomed to repeat it. But what of those who embrace history, who seek to draw wisdom from the annals of time? The blood that once divided us now has the power to unite, to teach us lessons of compassion and shared humanity.

Consider this: In a world where ancestry can be quantified, where the very blueprint of life is distilled into percentages and pie charts, what becomes of our sense of self? Are we not more than the sum of our parts, more than the echoes of our forebears' choices?

In the silence between heartbeats, listen. Therein lies the rhythm of generations past, the cadence of stories untold. Each pulse is a word, each throb a sentence in the ongoing narrative of blood and ancestry.

Therein lies the challenge. To see beyond the crimson hue of blood, to read between the lines of genetic code, and find the common ground that makes us all irrevocably human.

And so, dear reader, we embark on this journey together. Through the pages of this book, we will explore the intricate dance of DNA and destiny, unravelling the threads of our shared tapestry. With each chapter, we will uncover the truth that blood, in all its complexity, binds us beyond race, beyond time. It is the lifeblood of our shared story, the essence that courses through the annals of history into the heart of our very being.

Blood and ancestry. They are not mere markers of where we have been, but signposts to where we are going. Let us tread this path with open hearts and minds, for it is in understanding our past that we can truly forge our future.

Blood Ties in Mythology

Beneath an indigo sky, the villagers gathered, eyes shimmering with the reflection of a bonfire that cracked and hissed, as if whispering ancient secrets. Among them stood a figure, draped in the garb of an elder, yet her eyes mirrored the wisdom of ages far beyond her years. Her name was Makeda, known to the people as the keeper of stories, the bridge between the now and the ever after.

Makeda's voice rose above the fire's roar, and the crowd leaned in, their hearts syncing to the rhythm of her tale. This night, she spoke of Aido-Hwedo, the cosmic serpent whose blood gave birth to the rivers and whose body shaped the mountains. In her words, the serpent slithered through the minds of her audience, majestic and terrifying.

Her narrative spun through the creation of the world; a tapestry of life woven from Aido-Hwedo's blood. The villagers felt the serpent's power, the life force pulsating in the veins of the earth. Makeda's tale was not merely a recounting;

it was an invocation, a call to the primal energy that connects all beings.

As the story unfolded, Makeda's gaze held a flicker of something deeper, a knowledge that this was more than myth. It was a reminder that blood carried the essence of life, a sacred fluid holding mysteries no science could fully unveil. She spoke of warriors anointed in the blood of dragons, claiming strength from the very beasts they defeated, of maidens who communed with the spirits of their ancestors through ritual and blood.

The unexpected journey came when Makeda revealed the serpent's vulnerability—its reliance on a singular, sacred place where it could rest. It was a vulnerability shared by all beings, the need for sanctuary, for connection, for ties that bound deeper than the flesh.

In the silence that followed the tale's end, the villagers found a universal truth within themselves: life's blood ties us not to fate, but to each other, to the shared pulse of existence. Makeda promised them wisdom, insights into the strength of their own blood, the power of their connections, the understanding that within them flowed the same life force that carved rivers and sculpted peaks.

To the reader, Makeda's story is a gateway into the profound realisation that blood is more than a biological truth; it is a symbol of our collective strength, a force that transcends the boundaries we create. The myths of old knew this power well, celebrating blood as the vessel of life and the currency of the soul.

In the pages to come, we will journey through these stories, seeking the wisdom they hold. What can the rites of blood in Norse sagas teach us about sacrifice and kinship? How does the vampiric lore of Eastern Europe speak to the fear and fascination with life's essence?

Will you not find, dear reader, that within each legend lies a thread of truth, a connection to the very beat of your heart? The myths are not mere stories; they are reflections of our deepest selves, mirrors of the blood that defines and unites us.

As we delve deeper, we will encounter the chimeric creatures of Greek mythology, born of blood spilled by the gods, and explore the Egyptian belief in blood as the divine fluid of pharaohs. Let us question together: What power lies in the stories of blood spilled, blood shared, and blood sanctified?

Each myth, each rite, each drop of blood shed in the name of power or piety is a piece of the puzzle of our existence. Through these tales, you will come to see that blood ties are not chains that bind us to a predetermined destiny but rather the ties that, in their infinite complexity, connect us to each other and to the timeless dance of life and death.

Remember, as you turn the pages, to ponder the significance of blood in your own life. How do the stories of your ancestors shape your understanding of self? Are you not part of a narrative that stretches back to the dawn of time, a narrative written in the blood of those who came before?

Let us then embark on this odyssey through myth and memory, seeking the echoes of ancient wisdom that resonate within our modern veins. For in the blood lies a power beyond measure, a tie that binds us, not just beyond race, but to the very essence of what it means to be alive.

Transcending Blood: Beyond the Physical

Embark on an odyssey that promises to redefine the boundaries of your understanding, an adventure that will not only illuminate the path to a deeper self-awareness but also unravel the enigmatic ties that bind us beyond the corporeal. In "Transcending Blood: Beyond the Physical," I invite you to join me on a transformative journey, one that will reveal the profound impact of the metaphysical connections through blood, connections that surpass the mere tangible.

As a medical science student, I have been deeply rooted in the empirical. Yet, it is in the synthesis of science and the spiritual that the true essence of blood's significance emerges. In these pages, you will uncover the methodologies that bridge the gap between what is seen under the microscope and what is felt within the soul. Through a tapestry of cultural insights, historical perspectives, and personal narratives, we will explore the ancient belief systems that have revered blood as the cradle of the spirit and the modern interpretations that seek to harmonise these beliefs with contemporary understanding.

Some may question the validity of such a quest. How can one assert the existence of a soul within the bloodstream, or claim a metaphysical connection when science dictates otherwise? To this scepticism, I offer an invitation to consider the limitations of scientific inquiry in its current form. Science, in its noble pursuit of knowledge, often neglects the subjective, the intangible experiences that define our humanity. It is within this realm that "Transcending Blood" operates, embracing the mystery and acknowledging the spaces between what is known and what is felt.

Picture yourself standing at the threshold of revelation, ready to explore the depths of your own being through the prism of blood's ancient and enduring legacy. Imagine discovering a kinship that transcends ethnicity, nationality, and even species—a universal bond that is as much a part of you as the blood coursing through your veins.

With each chapter, you will feel the stirrings of a connection that defies explanation, a sense of belonging to something greater than the sum of your genetic material. The stories contained within these pages will not only challenge you to look beyond the physical but also lead you to question the very foundations of identity and kinship.

By committing to this journey, you are not merely reading a book; you are engaging in an act of self-discovery that has the power to alter your perception of life itself, is not just a collection of words; it is a gateway to a new understanding, a manifesto for those who seek to embrace the full spectrum of connection that blood represents.

Delve into the heart of this exploration with an open mind and emerge with a heart enlivened by the possibilities of what lies beyond. The commitment you make now to journey through these pages is a step toward a profound awakening—an awakening to the ties that bind us not just in flesh and bone, but in spirit and essence.

With vivid narratives that paint the ancient rituals in hues as vibrant as life itself, your imagination will be captured by the enduring power of blood to speak to our souls. Through direct questions posed to your innermost self, you will engage in a dialogue that transcends the written word, becoming an active participant in the unravelling of this age-old mystery.

You will encounter language that is both accessible and evocative, each sentence crafted to resonate with the rhythm of your own pulse. Quotes from sages and scientists alike will pepper the text, offering diverse perspectives that enrich the tapestry of this discourse. And through the power of anecdote and personal testimony, you will witness the living truth of the connections that defy explanation, that hint at a reality beyond the flesh.

In "Transcending Blood: Beyond the Physical," the profound awaits. Be ready to embrace a vision of humanity that is as boundless as the soul, as eternal as the lifeblood that links us all. Welcome, dear reader, to a transformative experience that transcends the physical, into the heart of what it means to be truly connected. Welcome to a journey beyond race, beyond the tangible—welcome to the realm of "Blood Ties: Beyond Race.

Blood Type Stereotypes

In a world that prides itself on scientific advancement and cultural enlightenment, one might assume that we have moved beyond the superstitions and folklore of our ancestors. Yet, beneath the surface of modern society simmers an undercurrent of stereotypes not based on race or creed, but on something as fundamental to our biology as blood. It's the kind of problem that seems almost too bizarre for the enlightened age we claim to inhabit, a problem that whispers of ancient times and yet thrives in the boardrooms and bedrooms of the twenty-first century.

This is the odd and often overlooked world of blood type stereotypes, a phenomenon where the very liquid that sustains us becomes the basis of broad assumptions about personality, compatibility, and destiny. It's a peculiar notion that one's blood type, an A, B, AB, or O, might somehow dictate their disposition and fate. But the implications of such beliefs are no laughing matter—they shape social dynamics, employment opportunities, and even the search for love.

Picture an elementary school playground in a bustling metropolis in East Asia. Children, not more than ten years old, are already whispering about who is the most trustworthy or daring based on the letters that define their blood. Consider the job postings that subtly suggest a preference for a specific blood type, hinting at the kind of personality traits they expect from their candidates. Imagine the dating profiles that list blood type alongside hobbies and zodiac signs, as if it were a crucial piece of the romantic puzzle.

How did we come to this curious juncture in our human journey? At the heart of this conundrum lies a history woven with pseudoscience and cultural narrative, a history that has imbued the types of blood coursing through our veins with meanings and biases that have little to do with their biological functions. The influence of this belief system is far-reaching, affecting individuals' self-perception and social interactions in tangible ways.

Consider the story of Yumi, a vibrant young professional from Seoul, who was denied a promotion because her blood type B was associated with being too free-spirited and unreliable for a leadership position. Or Ken, a meticulous and dedicated Tokyo office worker, who was overlooked for a creative project because his blood type 'A' was deemed too conservative and risk averse. These are not isolated anecdotes; they are representative of a larger issue that questions our commitment to individuality and meritocracy.

The stakes, as it turns out, are high. When we allow such stereotypes to persist unchallenged, we risk perpetuating a form of discrimination that is as insidious as it is overlooked. We risk the integrity of our social fabric, woven with the threads of fairness and justice, by allowing a mythological understanding of blood to dictate the opportunities and relationships afforded to individuals.

Yet there is hope on the horizon. This book will not only dissect the origins and manifestations of blood type stereotypes but will also chart a course towards dispelling these myths. Through a blend of scientific evidence, sociological study, and personal narratives, we will unravel the complex web of beliefs and biases that have clung to the very lifeblood of our existence.

As we journey together through these pages, you will be introduced to experts who challenge the status quo, individuals who have overcome the stigma attached to their blood type, and societies that are beginning to question the place of such stereotypes in a modern, rational world.

Is it not time to ask ourselves why we cling to these archaic notions? What do we gain from believing that the essence of who we are can be summed up by the type of blood that flows through our veins? And, more importantly, what do we lose?

In a world where individuality is celebrated, where the content of one's character holds more weight than the colour of their skin or the origin of their birth, the perpetuation of blood type stereotypes is an anomaly that demands our attention. It is a relic that belongs in the past, yet it persists, a shadow that follows us, unacknowledged, into the future.

By the time you turn the last page of this book, you will not only understand the absurdity of such stereotypes but also be equipped with the knowledge and insight to challenge and change them. Together, we can move towards a society that sees beyond the type of blood that binds us to one that recognises the shared humanity that truly defines us.

In the chapters that follow, we will embark on a journey to dismantle the myths of blood type stereotypes. We will explore their historical roots, their prevalence in various cultures around the globe, and the psychological underpinnings that allow such beliefs to flourish. With each step forward, we will shine a light on the truth that lies beyond the labels, a truth that liberates us from the confines of misguided beliefs and brings us closer to the essence of what it means to be part of the human family.

Blood ties us together, but it is the heart and the mind that truly define us. Let us forge ahead, beyond race, beyond stereotypes, and into a realm where every individual is seen for who they truly are, not the blood type that runs in their veins.

Blood as a Social Construct

In the intricate tapestry of human society, the threads of blood and race have been interwoven to create patterns of division and unity, often with far-reaching consequences. Blood, a vital fluid coursing through our veins, and race, a complex constellation of physical traits and cultural constructs, have both served as potent symbols for social categorisation. Their significance lies not only in the biological realm but also in the intricate choreography of human interactions and the silent, sometimes invisible, barriers they erect.

This exploration seeks to dissect the concept of blood as a foundation for social grouping, juxtaposing it against the notion of race, to unveil the subtle, yet powerful, forces that shape our perceptions and experiences. What can the comparison of these two subjects reveal about the ways we construct identity and the implications for our collective journey?

At the heart of this inquiry is a quest to understand the rationale behind using blood as a marker of social difference. Much like race, the idea of blood as a signifier of belonging or exclusion has a storied history, punctuated by both scientific inquiry and mythmaking. By setting the benchmarks for comparison, we delve into the shared attributes and the divergent paths these concepts have carved in the landscape of human association.

Consider the way in which both blood and race have been utilised to forge communities. The similarity lies in their use as a basis for forming in-groups and out-groups, whether it be through bloodline kinship or racial affinity. The bonds of blood, from familial ties to the notion of 'pure' lineages, have historically dictated one's status, inheritance rights, and social circles. Similarly, race has been employed to demarcate the boundaries of social inclusion, often with a focus on phenotype as a proxy for a shared history and culture.

Yet, it is in their differences that we find the nuanced strands of our social fabric. Blood, in its literal sense, is a universal human trait, undiscriminating and unifying. Yet, the metaphorical use of blood has carved out a notion of heredity that is both exclusive and restrictive. In contrast, race, while ostensibly based on physical characteristics, is largely a social invention, its definitions malleable and subject to the whims of societal norms and power dynamics.

The analysis of these comparisons uncovers a profound insight: both blood and race are social constructs, their meanings and implications shaped by human perception and societal conventions. The broader implications of this realisation are significant. By understanding the artificiality of these divisions, we can begin to unravel the systemic biases that have been built upon them and work towards a more inclusive society.

The relevance of these comparisons to contemporary realities is undeniable. In an era where genetic testing can reveal a person's ancestry, the concept of blood as a basis for identity is being both reinforced and challenged. At the same time, the ongoing discourse on race, with movements advocating for racial equality and justice, highlights the continued impact of racial categorisation on people's lives.

What does it mean, then, to live in a world where the very essence of our being – our blood – has been co-opted as a marker of division? Is it not time to confront these constructs that persist, despite their shaky foundations? Are we ready to envision a society where the content of our character, our actions, and our choices speak louder than the silent markers of blood or race?

In the subsequent chapters, we will delve deeper into the consequences of these social constructs, exploring their manifestation in various cultural contexts and their effects on individual identity. We will meet individuals who have transcended the limitations imposed by blood and race, and we will examine the role of science and education in dismantling these outdated beliefs.

The journey ahead is one of enlightenment and transformation—a path that requires us to question the status quo and embrace a future where our common humanity supersedes the artificial divides of blood and race. As we forge this new path, let us remember it is not the blood in our veins that defines us, but the shared blood of humanity that unites us. Together, we can redefine the ties that bind, moving beyond the confines of blood and race to a place where our shared human experience is the only tie that truly matters.

CHAPTER 3

The Essence of Blood

Embarking on this enlightening journey, one must first grasp the lexicon that defines the very essence of our investigation: the ABO blood group system. Mastery of these terms is pivotal to navigate the intricate pathways that link our shared biology to the broader expanse of human connections.

The terms to be mapped out in our intellectual odyssey include 'antigen,' 'antibody,' 'genotype,' 'phenotype,' 'universal donor,' 'universal recipient,' and the 'Rh factor.' Each of these holds a key to unlocking the mysteries of blood types and their profound implications on human interaction and medical science.

Antigens are substances that the body recognises as foreign, stimulating the immune system to produce antibodies. Specifically, in the context of blood types, they are protein molecules found on the surface of red blood cells. Each blood type has a unique antigen profile, distinguishing it from others.

Antibodies, on the other hand, are the immune system's defenders. They seek out antigens that do not match the body's own, marking them for destruction. Within the bloodstream, the presence of incompatible antibodies can lead to agglutination, a potentially fatal clumping of cells.

Genotype refers to the genetic makeup of an individual that determines their blood type. It is the combination of alleles inherited from one's parents. The genotype remains unseen, yet it is the blueprint for the phenotype—the observable characteristics, such as blood type, determined by the genotype.

The universal donor, a term synonymous with Blood Type O negative, is one whose red blood cells can be transfused into individuals of any other blood type without fear of adverse reaction due to the absence of A or B antigens. Conversely, the universal recipient—Type AB positive—can accept blood from any donor, as their body does not produce antibodies against A, B, or Rh antigens.

The Rh factor, an additional antigenic system, further categorises blood types into positive or negative. This factor is named after the Rhesus monkey, in which it was first discovered. The presence or absence of this antigen adds another layer of complexity to blood compatibility.

How often does one reflect on the silent symphony of cellular interactions within their own veins? Do you not wonder at the microscopic dance of antigens and antibodies that occurs each time blood is transfused? These players in the drama of life and death are as crucial to our survival as the more visible traits that define our identities.

Imagine a scenario where a life hangs in the balance, and only a precise understanding of these terms can guide the hands of healers. Picture the expectant mother whose unborn child's life depends on the Rh factor compatibility with her own. Or consider the soldier on the battlefield receiving the gift of life from a universal donor, whose identity is a mystery, but whose blood is universally accepted.

In the grand tapestry of human existence, these terms are not mere academic jargon. They are the threads that weave together stories of survival, hope, and connection. They shape the narratives of who we are in the most profound ways—beyond the superficialities that have too often divided us.

As we move forward in our exploration, we shall delve ever deeper into the history and implications of these terms. We will examine how the discovery of the ABO blood group system by Karl Landsteiner in 1901 revolutionised medicine and paved the way for safe blood transfusions. We will uncover the rich tapestry of genetics that dictates blood type inheritance. And we shall confront the ethical and social questions that arise when the essence of our physical beings—our blood—becomes a commodity that can be both lifesaving and, in the wrong hands, life-threatening.

This understanding is not an academic exercise; it is a foundation upon which we can build a more compassionate and equitable society. It is a beacon that guides us through the murky waters of ignorance and prejudice, towards a shore where the content of our character and the compatibility of our blood eclipse the outdated constructs of race.

And so, without further ado, let us embark upon this voyage of discovery. Let us commit these terms to memory and weave them into our understanding of the world. For in the intricate dance of antigens and antibodies, in the silent language of genotypes and phenotypes, lies a story of human connection waiting to be told.

The History of Blood Typology

In the shadowy annals of medical history, the tale of blood typing emerges like a beacon, guiding us through a labyrinth of scientific breakthroughs that have shaped the very fabric of healthcare. It is a story punctuated by moments of genius, serendipity, and profound human insight.

At the heart of this voyage is a simple question: what makes our blood unique, and why does it matter? But the answers to this question are anything but simple, as they unravel in a complex web of biological discovery and human drama.

The genesis of blood typing lies shrouded in the mists of time, long before the first microscope revealed the cellular wonders of our blood. Ancient civilisations were not privy to the secrets hidden within our veins, but they were acutely aware of blood's significance, be it in rituals, medicine, or the arts of war.

Understanding the intricate framework of blood types is tantamount to appreciating the rich tapestry of human biology. The very essence of our survival flows through our veins, a crimson river teeming with secrets waiting to be unlocked. Diving into the depths of hematologic science, one discovers that blood is not merely a substance but a storyteller, holding the narrative of life itself.

Embark on a journey through the lifeblood of humanity, where the alphabet of existence is etched in every drop. It is here, in the serenity of science, that we encounter the key players of our story: A, B, AB, and O. These characters, though simple in their nomenclature, are the architects of compatibility and the guardians of genetic legacy.

As we unfurl the scroll of understanding, anticipate an odyssey through the world of antigens and antibodies, donors and recipients, universal lifelines, and the rarest of blood's nectar. Let us, with meticulous care and unbridled curiosity, explore the wonder that is blood typology.

In the realm of antigens, tiny sentinels standing guard on the surface of blood cells, the plot thickens. They are not mere decorations but vital indicators of identity, signalling to the immune system whether cells are friend or foe. The A and B antigens are the protagonists of this tale, with O lacking these markers entirely, and AB proudly presenting both. Their presence or absence on the red cell's surface dictates the blood type, etching a person's uniqueness into the annals of biological diversity.

Now, consider the antibodies, vigilant defenders in the plasma, always on the lookout for foreign invaders. They are the sworn enemies of antigens not recognised as self. A person with Type A blood produces anti-B antibodies, while Type B blood harbors anti-A antibodies, and Type AB does not produce these antibodies at all. Type O, the chameleon, is the universal donor, cherished for its lack of A and B antigens, but its plasma teems with both anti-A and anti-B antibodies, a dichotomy of generosity and defence.

The Rh factor, an additional layer of complexity, is a character in its own right. Often represented as a simple plus or minus sign, it is anything but simple. Rh positive denotes the presence of another antigen, the D antigen, while Rh negative lacks this component. This distinction is crucial, particularly in pregnancy, where the interplay between maternal and foetal blood types can unfold as harmony or discord, depending on the compatibility of their Rh factors.

What, then, of the practicalities of this knowledge? Transfusion medicine turns to blood types to choreograph the dance between donor and recipient, ensuring that this life-sustaining ritual is performed with precision and care. It is a delicate balance, one that can mean the difference between life and death.

Have you ever wondered why certain blood types are sought after during emergencies? The O negative blood type, often exalted as the universal donor of red cells, is a beacon of hope in the most dire of circumstances. AB positive, on the other hand, reigns as the universal recipient, its open arms accepting red cells from any blood type. These roles are reversed when considering plasma: AB becomes the universal plasma donor, and O the universal recipient. Such is the paradox of blood typology.

Yet, beyond the clinical, the concept of blood type stretches into the fabric of society itself. In some cultures, it permeates personality theory, with attributes ascribed to each blood type as if written in the stars. A curious notion, indeed, but one that captivates the minds of many.

The river of life flows ever onward, and as it meanders through the landscape of our understanding, it nourishes the fields of knowledge. Blood typology, once a whisper in the corridors of science, now speaks with the clarity of truth, echoing the message that within our veins runs the potential for unity.

In the absence of a conclusion, let this be an invitation to ponder. Reflect on the marvel of our shared biology, the common thread that weaves through every human heart. There is no question to pose, no final statement to make, only the silent acknowledgment that within us all runs the same life-giving force, a force more binding than any construct of race.

Blood Transfusion Taboos and Triumphs

Blood transfusions are a case in which the old age that "opposites attract" cannot be proven true. Although it is generally accepted that opposites attract, there are several exceptions to this rule when it comes to blood. Blood transfusion, a practice dating back centuries, stands as a testament to the incredible strides made in medical science. From its early, often mystical origins to the sophisticated procedures of today, the journey of blood transfusion is one rich with innovation, ethical quandaries, and cultural variations that have shaped its current state.

At the dawn of this practice, the understanding of blood's significance was both revered and feared. Ancient civilisations, from the Mayans to the Egyptians, held blood as a symbol of life and vitality, imbuing it with spiritual significance. It wasn't until the 17th century that the earliest recorded attempts at blood transfusion took place.

Have you ever given any thought to when they would have gotten such an odd name? The recognition of diverse blood types among individuals was not understood until the eighteenth century. Fatalities were a common occurrence in blood transfusions, especially in cases when the donor and recipient were closely related or of the same racial background. This was mostly due to the recipient's immune system encountering the transfused blood as alien and initiating an immunological response against it.

The practise of blood transfusion has a lengthy historical background that may be traced back to the 17th century, along with the ground-breaking discovery of the circulatory system by William Harvey. In 1658, Jan Swammerdam employed a microscope to observe red blood cells. The initial instances of blood transfusions were conducted on dogs, as Richard Lower, an English physician, successfully illustrated the viability of sustaining a dog's life by the infusion of blood obtained from other dogs.

Unfortunately, blood transfusions for humans were so difficult and complicated. Back in the day, blood transfusions were extremely dangerous because there was no understanding of blood group. In 1818, James Blundell performed the first successful human blood transfusion, saving a woman who was haemorrhaging from childbirth.

However, blood transfusions were unsuccessful in some cases. Many people went into shock and died as a result of blood transfusions. Due to this reason, the transfusion of blood substitutes was one method used by some scientists to try to stop the negative effects of blood transfusions.

In 1854, during a cholera epidemic in Canada, attempts were made to transfuse cow's milk. Dr James Bovell and Edwin Hodder began administering intravenous milk transfusions because they thought that milk's fat molecules could be converted into white blood cells, which were an immature form of red blood cells.

Of course, this assumption was completely wrong, but they were successful with one sick person who responded well to the transfusions. But after milk was infused into the veins of two additional patients, they both passed away.

After the cholera outbreak, these experiments were suspended in Canada but restarted in New York City a few years later. Dr Joseph Howe transfused patients with terminal tuberculosis this time around using goat's milk. All the patients experienced chest pain and nystagmus (shaking eye movements), and again, all patients passed away shortly after receiving the transfusion.

Even though milk transfusions were not clearly beneficial, they continued in the late 1880s because using blood was not recommended because of its strong tendency to coagulation. The use of milk transfusions steadily declined as more patients died suddenly from milk transfusions.

In 1880, isotonic saline solution was developed, and the use of milk completely faded away in favour of the new, safe saline solution. The understanding of various blood groups and compatibility would not be fully understood until the 20th century, halting the revival of blood transfusions.

Figure 1. *Karl Landsteiner*

Photo from The Nobel Prise Foundation.

In 1900, at the University of Vienna, Karl Landsteiner, an Austrian-born American biologist, made his first ground-breaking discovery about why certain blood transfusions were successful while others could be fatal.

When Landsteiner combined the red blood cells and serum from each member of his staff, he found the ABO blood type system. He showed how different people's serum might cause their red blood cells to aggregate. He discovered three varieties from these early studies, which he labelled A, B, and C (C was later to be renamed O for the German "Ohne," which means "without," or "Zero," which is "null" in English).

A year later, researchers identified the fourth, less common blood group, AB. For his work, Landsteiner won the Nobel Prise in physiology or medicine in 1930. However, for the first time in medical history, the crossmatch of blood was performed in 1907. Even though the compatibility problems had been resolved, blood still had a propensity to clot during storage, which was a problem in the early days of blood transfusion.

In 1914, anticoagulants like sodium citrate were created, enabling blood to be stored for long periods of time. Then another issue arises: the Rh antigen. In 1940, the Rh factor was discovered, which enabled medical professionals to fully comprehend the problems relating to compatibility between blood donors and recipients. Shortly after, the American government launched its first comprehensive national blood collection programme.

Figure 2. *Photo from Dr. Blundell's article in the June 13, 1829, issue of The Lancet,*

"Observations on Transfusion of Blood."

The visual timeline of blood transfusion's history is as stark as it is enlightening. Diagrams of early transfusion apparatuses, photographs of the pioneers in the field, and charts of blood type distributions across populations tell a story that words alone cannot fully capture. These images serve as silent witnesses to the challenges and triumphs that have marked the evolution of transfusion medicine.

Cultural and regional variations in blood transfusion practices have been notable. In some cultures, taboos surrounding the mixing of blood meant that transfusions were shunned, while in others, the collective good of the community overrode individual hesitations. In the 20th century, war became a catalyst for innovation in blood transfusion, with the establishment of blood banks and the use of mobile blood donation units. These developments were not uniform across the globe but were instead shaped by the unique needs and resources of different nations.

The modern era has brought with it an array of interpretations and adaptations of blood transfusion practices. The rise of autologous blood donation, where patients donate their own blood in anticipation of surgery, and the advent of blood substitutes highlight the ongoing quest to improve and diversify transfusion medicine. Yet, despite these advances, the ethical challenges remain as pressing as ever.

How can we reconcile the desperate need for blood with the imperative to protect donor autonomy and well-being? This question echoes through the corridors of hospitals and legislative chambers alike. The controversy surrounding paid blood donations brings to light the delicate balance between incentivising donation and the risks of exploiting vulnerable populations. Similarly, the implementation of stringent donor eligibility criteria, while crucial for patient safety, has sparked debates over discrimination, particularly in the context of restrictions based on sexual orientation.

A turning point in the ethical discourse came with the HIV/AIDS crisis of the 1980s. The realization that tainted blood could transmit deadly viruses led to a seismic shift in blood screening practices and donor policies. The introduction of rigorous testing protocols for infectious diseases such as HIV, Hepatitis B and C, and West Nile virus has since become a non-negotiable aspect of transfusion medicine.

As we contemplate the future of blood transfusion, are we prepared to confront the ethical dilemmas that will inevitably arise? Will technology and scientific discovery continue to outpace the frameworks we've constructed to govern this life-saving practice?

In the end, the story of blood transfusion is a mosaic of human endeavour, a narrative that weaves together the threads of scientific discovery, ethical reflection, and the unceasing quest for better ways to save lives. The blood that courses through our veins is more than a biological necessity; it is a symbol of our shared humanity, a reminder that the ties that bind us go beyond race, transcending individual differences to unite us in the common cause of preserving life.

Major Milestones: Chronological Sequence

Landsteiner's revelation was but the first chapter in an ongoing saga. In 1902, his colleagues Alfred von DE Castello and

Adriano Strule added the AB group to this classification, completing the ABO blood group system. The plot thickened in 1940 when Landsteiner and Alexander S. Wiener discovered the Rh factor, introducing a new dimension to blood compatibility.

The interwar period was a fertile ground for the growth of blood type science, with the establishment of blood banks and the refinement of storage and transfusion techniques. The scene was set for a revolution in medical practice that would save countless lives on both the battlefield and the home front.

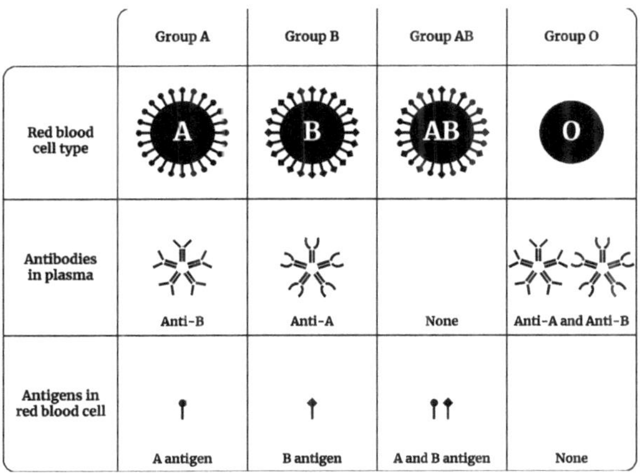

Figure 3. *The presence of ABO blood group antigens on Red Blood Cells that plays a significant role in determining an individual's blood type.*

The plot of blood typing is not without its twists and turns. Across the globe, the prevalence of blood groups varies, painting a diverse picture of humanity. For instance, the B blood type is more common in Asia than in Europe, and certain African populations have a higher frequency of the Rh-negative factor. These variations form a rich tapestry, revealing the interplay between genetics, evolution, and migration.

Fast forward to the present, and the narrative of blood typing continues to evolve. Advances in molecular biology have unveiled the genetic underpinnings of the ABO and Rh systems, while cutting-edge technologies like blood genotyping promise to refine our understanding even further.

The discovery of additional blood groups—over 600 distinct antigens to date—has turned our tale into an ever-expanding universe of knowledge, challenging scientists, and healthcare professionals to continually adapt their practices.

Yet, the journey of blood typing is not without its perils. The very knowledge that empowers us to save lives also bears the burden of ethical conundrums and controversies. The spectre of blood type discrimination, particularly in job recruitment and social relationships in some cultures, raises profound questions about the misuse of scientific knowledge.

Moreover, the advent of synthetic blood and the exploration of universal blood substitutes represent turning points that may one day redefine the very essence of blood typing. These innovations beckon us to a future where the constraints of blood group compatibility are but a distant memory.

As the story of blood typing unfolds, one cannot help but marvel at its intricacy and its impact on humanity. From the early days of Landsteiner's discovery to the high-tech labs of today, this saga of scientific endeavour is a testament to the human spirit's quest for understanding.

The intricate ballet of blood types has indeed proven to be more consequential than the superficial divisions of race. It is a narrative not of division, but of connection—a universal language written in the very lifeblood of humanity.

So, let us ask ourselves, what new chapters will we write in the history of blood typing? How will the knowledge we gain today shape the healthcare of tomorrow? These questions loom large on the horizon, beckoning us to continue our quest for knowledge and understanding.

Haemoglobin and Oxygen: The Lifeline Duo

The essence of vitality, the quintessence of existence, pulses through the highways of our circulatory system. This lifeline is nothing less than the sacred alliance of haemoglobin and oxygen, a duo that wields the power to sustain life as we know it. In the grand theatre of our bodies, this coupling performs a ballet so critical that its disruption spells catastrophe for the organism it nurtures.

Yet, what happens when this bond is threatened? With every breath, we are reminded of the delicate balance that must be maintained, for the consequences of failure are dire and far-reaching.

Breath itself is the invisible siren call that beckons oxygen to enter our being. Once inside, it embarks on an odyssey, navigating the winding paths of our lungs to find its partner in survival—haemoglobin. This iron-bearing protein resides within the red blood cells, waiting with open arms to embrace the oxygen molecules. Together, they form an alliance, a partnership that fuels the metabolic fires of every cell.

But consider this: what if haemoglobin falters in its duty? What if the embrace is loosened, the bond weakened? The result is a cascade of potential devastation. Tissues starved of oxygen languish and wither, organs struggle to perform their vital functions, and the very fabric of life begins to unravel.

Such is the plight of those who suffer from anaemia, a condition marked by insufficient haemoglobin and a consequent lack of oxygen transport. The implications are profound—an insidious fatigue that permeates every aspect of life, a vulnerability to infections, and a dampening of the bright flame of vitality.

Yet, the solution to this conundrum lies within reach, rooted in both nature and innovation. Our understanding of the biological mechanisms that govern the synthesis of haemoglobin has burgeoned, illuminating pathways to intervention. Iron, the core of haemoglobin's oxygen-binding ability, becomes a focal point of our strategy. Iron-rich diets, supplements, and treatments that address the underlying causes of anaemia can restore the delicate balance within the bloodstream.

But how do we implement this knowledge? The answer begins with awareness and education, informing those at risk of the signs and symptoms to watch for, the importance of nutrition, and the availability of medical evaluation. Healthcare systems must prioritise accessible testing for anaemia and provide the necessary resources for treatment.

Evidence of the efficacy of such interventions is not elusive. Studies have shown that when individuals with anaemia receive appropriate iron supplementation, their haemoglobin levels rebound, and with it, their vitality. The fog of fatigue lifts, cognitive function sharpens, and the body's resilience is fortified.

And yet, a singular solution is not a panacea for all. Alternative therapies, such as blood transfusions in acute cases or the administration of erythropoietin to stimulate red blood cell production, may be warranted. Each case demands a bespoke approach, tailored to the unique tapestry of the individual's physiology.

What of those who walk the earth unaware of the silent war waging within their veins? How often do we overlook the subtle whispers of our bodies, dismissing them as the mere ebb and flow of life's tides? It is here that we must pose the question: Are we listening to the symphony of signals that our bodies orchestrate, or do we await the crescendo of crisis?

Our charge is clear. We must not only listen but respond with the full complement of our knowledge and compassion. We must weave a net of prevention and treatment, a safety mesh that catches those who fall into the chasm of ill health.

In this pursuit, we invoke the rhythm and cadence of life—short, sharp breaths when urgency strikes, the steady inhalation of resolve, and the long exhale of relief as balance is restored. Through the vivid imagery of health regained, we paint a portrait of hope, a testament to the tenacity of the human spirit and the ingenuity of medical science.

Let us then, with simple language and a reverence for life's complexity, champion the cause of haemoglobin and oxygen, the lifeline duo. For in their dance lies the very essence of our being, a melody that resonates with the force of life itself.

Blood Disorders: A Glimpse into Pathology

In the labyrinth of the human body, the blood serves as a river of life, carrying not only the essential oxygen and nutrients but also the tales of our health, whispered through its currents. As the sun dips below the horizon, casting a crimson hue that mirrors the lifeblood within us, we embark on an exploration of the pathologies that can disrupt this vital flow. Tonight, we uncover the veiled mysteries of blood disorders, examining how these silent saboteurs can overturn the delicate equilibrium of our existence.

As the moon takes its watchful place in the night sky, let us introduce our main players in this nocturnal narrative: the platelets, the white blood cells, and the plasma that houses a myriad of proteins and factors, each with a role more critical than the last. In the shadows, they perform their duties unseen, yet their importance cannot be overstated.

Imagine, if you will, a bustling cityscape—the veins and arteries are the streets, the blood cells are the citizens, and the plasma, the very air they breathe. In this metropolis of metabolism, a challenge arises: a disorder known as haemophilia. This genetic anomaly whispers through generations, marking its presence not by what it brings, but by what it takes away—the ability to clot blood effectively.

The approach to confronting this challenge is multifaceted. It begins with the identification of the missing clotting factors, the proteins that, when summoned, converge to form a dam, stopping the flow of blood from a wound. With modern medicine's quiver, we introduce replacement therapy, providing the missing factors through intravenous infusions. The results, a shield against uncontrolled bleeding, have transformed lives, allowing those affected to live with a semblance of normalcy.

But let us analyse and reflect for a moment. While treatment offers a reprieve, the spectre of complications looms—antibodies may form against the very factors meant to heal, and the risk of blood-borne infections from transfusions, though reduced, is never zero.

Yet, to connect this to our larger narrative, haemophilia is but one actor on the stage of blood disorders. There are others, like the enigmatic sickle cell disease, where the red blood cells, deformed into crescents, journey through the bloodstream, causing blockages and pain. Or leukemia, where white blood cells proliferate wildly, crowding out their healthy counterparts.

How then do we transition from the contemplation of these maladies to a broader engagement with our health? What questions should we ask when the whispers of our body suggest a deeper malaise? Could it be that our fatigue, our bruises, our recurrent infections, are not merely stray notes but a symphony playing to the tune of a blood disorder?

In pursuit of answers, we must advocate for routine blood tests, those silent sentinels that guard against the unseen. We must educate, not just in the hallowed halls of academia, but in the public square, on the signs and symptoms that herald these conditions.

We stand at the threshold of a new dawn in blood pathology, where the once-impenetrable darkness gives way to the light of understanding. As a medical science student, I've witnessed the power of knowledge to dispel fear and replace it with action.

Let us wield this knowledge with the precision of a surgeon's scalpel, cutting away ignorance and apathy. Let the stories of those who live with these conditions not be muffled by the clamour of the undisturbed but be amplified, a chorus calling for awareness and empathy.

The journey is long, and the night is but a part of it. The true measure of our progress lies not in the distance travelled but, in the steps, we take toward a brighter, healthier tomorrow. Will you walk this path with me, dear reader, towards a future where the ties of blood are not a burden but a bond that unites us in pursuit of well-being?

As the first light of dawn stretches across the sky, let us ponder this: what role can each of us play in the grand narrative of health? How can we contribute to the understanding and treatment of blood disorders that touch the lives of so many?

Let us continue to ask, to learn, and to care. For in the end, it is not merely our own blood that ties us to the race of humanity, but our collective commitment to transcend beyond our individual existence, reaching out to those in need, bound by the most profound of all ties—compassion and empathy.

Blood Types and Genetic Inheritance

The fabric of our genetic heritage is woven with threads of mystery, each strand representing the biological legacies passed down through generations. As we delve deeper into the science of genetics, particularly with regard to blood types, we uncover truths that transcend the superficial notions of race and ethnicity. This section of our journey explores how blood types are inherited, the role they play in our lives, and the myths they dispel about racial inheritance.

At the core of this exploration is the ABO blood group system, with its four primary blood types: A, B, AB, and O. The ABO blood type is inherited from our parents, each of whom contributes one of two ABO genes—A, B, or O—which combine to determine our blood type.

Consider, for instance, a child born to a mother with type A blood and a father with type B blood. This child could potentially inherit an A gene from the mother and a B gene from the father, resulting in type AB blood. Alternatively, the child could inherit an O gene from either parent, leading to either type A or type B blood. The genetic permutations are fascinating, and they underscore the fact that our blood type is not a mark of race but a mosaic of our parents' genetic contributions.

While the ABO system is the most recognised, other systems also play a role in genetic inheritance, such as the Rh factor, which classifies blood as either positive or negative. This, too, is an inherited trait, with positive being the dominant allele. The Rh factor becomes particularly important in pregnancy, as Rh incompatibility can lead to complications.

The idea that certain races are predisposed to specific blood types is a myth that has been debunked by genetic research. While it is true that certain blood types are more or less common in different populations around the world, this distribution is not aligned with the concept of race as it is socially constructed. Rather, it reflects patterns of human migration and adaptation to various environments over thousands of years.

To illustrate the point, let us consider the O blood type, which is more prevalent among Indigenous peoples of the Americas and Africa. This prevalence is not indicative of a racial trait but is likely a result of genetic drift and the founder effect, where a small population that separates from its parent population carries only a fraction of the genetic diversity.

Moreover, the distribution of blood types within any given population is not homogenous. For example, among African populations, there is a wide range of blood type distributions, refuting any simplistic racial categorisation. This diversity within groups further emphasises that blood type is not a racially exclusive trait.

When we look at the hard data, it's clear that blood types cut across racial lines. In some regions of the world, such as Sub-Saharan Africa, the distribution of the Rh-negative blood type is less than 1%, while among the Basque people of Europe, it can be as high as 35%. Such data challenge the notion that blood types are a marker of racial difference and instead point to a complex tapestry of human genetics shaped by evolutionary forces.

As we break down complex terms like "allele frequency" and "genetic drift," we find that these concepts are the keys to understanding the rich diversity of blood types. Allele frequency refers to how common an allele, or version of a gene, is in a population. Genetic drift is a mechanism of evolution that refers to random changes in allele frequencies, which can lead to significant differences over time.

In concluding this section, we must recognise the key take-aways: blood types are inherited through a fascinating interplay of genetics that does not correspond with the socially constructed categories of race. The understanding of blood types and their inheritance patterns is not only a testament to the complexity of human biology but also a powerful reminder of our shared humanity.

Despite the diverse tapestry of blood types that exist within and across populations, they reveal a story of human connection rather than division. As we move forward, let us continue to unravel the intricate genetic threads that bind us all, acknowledging that the true significance of blood types lies in their ability to save lives, not in their ability to separate us by race.

Universal Donors and Recipients

In the intricate web of transfusion medicine, a stark reality emerges; not all blood is created equal when it comes to the life-saving act of blood transfusion. Amidst the myriad of blood types, two stand out for their unique roles in this critical field: O negative, the universal donor, and AB positive, the universal recipient. The availability and use of these blood types carry profound implications for healthcare systems around the world.

Envision a world where an accident victim arrives at the emergency room, their life hanging by a thread, in dire need of a blood transfusion. The clock ticks ominously; there is no time to crossmatch blood types. This is where the critical role of O negative blood emerges. Devoid of A, B, or Rh antigens, O negative blood is a beacon of hope, a universal donor whose red blood cells can be transfused into any patient, regardless of their own blood type.

But what if this lifeline is in short supply? The consequences are dire: delayed surgeries, compromised emergency care, and a grim increase in preventable deaths. The scarcity of O negative donors, who constitute a mere 7% of the population, exacerbates this potential crisis. It's a tightrope walk between life and death, where the balance is maintained by the availability of this precious resource.

Now, consider the opposite end of the spectrum: AB positive individuals, the universal recipients. Their bodies accept all types of blood, an emblem of inclusivity in the world of transfusion. However, they represent only about 3% of the population, a small group with a mighty advantage in emergencies. But their rarity is double-edged; finding donors for them is not a challenge, yet their blood cannot be as widely shared with others.

So, how do we navigate this conundrum of supply and demand? The solution lies in a multi-faceted strategy. First, we must amplify public awareness about the critical need for O negative donors. Blood drives, media campaigns, and educational programs can illuminate the significance of this blood type and inspire a new generation of donors to step forward.

Secondly, hospitals and blood banks should implement meticulous inventory management, ensuring that O negative blood is reserved for true emergencies and cases where it is unequivocally necessary. This judicious use can prevent wasteful depletion of the universal blood supply.

In parallel, advances in medical research offer a beacon of hope. Scientists are working tirelessly to develop blood substitutes or methods to convert other blood types into O negative. While these solutions are still on the horizon, their potential to revolutionise transfusion medicine is immense.

Evidence of the success of these strategies can be found in countries with robust donor programs. For instance, in Sweden, a mobile app notifies donors when their blood has saved a life, reinforced the personal impact of their donation and encouraging ongoing commitment. Such innovative approaches cultivate a strong donor base, ensuring a steadier supply of universal blood types.

Moreover, there's room to consider alternatives. Could we, perhaps, develop synthetic blood products that perform the same oxygen-carrying functions as human blood? While still in the experimental stage, these could eventually provide a supplementary option in critical situations.

The gravity of this issue cannot be overstated. Each drop of O negative blood is a potential lifeline, and every AB positive individual is a testament to the remarkable adaptability of the human body. It is our collective responsibility to ensure that this knowledge translates into action. Can we afford to ignore the call for more universal donors? The answer must unequivocally be no.

The essence of our humanity is reflected in the blood coursing through our veins—a common thread that connects us all. In the grand mosaic of life, blood types matter not for their ability to divide, but for their unmatched power to unite us in the face of adversity.

As we advance in our understanding and stewardship of these universal donors and recipients, let us remember that the choices we make today can mean the difference between life and death tomorrow. The call to action is clear, and the time to act is now. Can we rise to the challenge? The future of transfusion medicine depends on it.

CHAPTER

4

Ethical Intricacies of Blood

In the heart of modern medicine, a life-sustaining elixir flows, its crimson waves carrying the very essence of survival. Blood, an element so vital, yet so taken for granted, has become the centre of an ethical maelstrom, one that questions the roots of our moral responsibility. As we delve into the depths of this issue, we must ask ourselves a crucial question: What are the moral considerations surrounding blood donation, and how do the policies that govern it affect us all?

Blood donation, the voluntary act of giving one's blood to assist others in medical need, has become a cornerstone of healthcare systems worldwide. Yet, beneath the surface of this altruistic endeavour lies a complex ethical dilemma. The primary issue at hand is the balance between donor autonomy and recipient safety, the struggle to ensure a sufficient blood supply while safeguarding the rights and well-being of those who give this precious resource.

If left unaddressed, the consequences of neglecting these ethical considerations could be dire. A shortage of blood can lead to a public health crisis, with hospitals scrambling to meet the needs of patients requiring transfusions. Conversely, exploitation or harm to donors can lead to a distrust of the medical system and a subsequent decline in donation rates.

The solution to this quandary rests on a multi-faceted approach that honours both the dignity of the donor and the needs of the recipient. Policies must be put in place that prioritise informed consent, non-coercion, and the safety of all parties involved. To this end, educational campaigns can be instrumental in demystifying the donation process, dispelling fears, and emphasising the importance of volunteerism in maintaining a robust blood supply.

Implementation of these policies would involve rigorous screening processes for potential donors, thorough testing of donated blood, and ongoing monitoring of the effects of donation on both donors and recipients. A feedback loop, where donors are appraised of the impact of their contributions, could foster a sense of community and ongoing engagement with the donation process.

Evidence of the efficacy of such solutions can be found in countries with high donation rates, where transparent practices and donor appreciation initiatives have led to a sustainable blood supply. In these cases, donors are often seen as heroes, a sentiment that encourages repeated and new donations alike.

While the proposed solution stands robust, alternative strategies merit consideration. For instance, paid donations, a controversial subject, may incentivise individuals to donate more frequently, but may also raise questions of coercion and the commodification of human body parts. Another approach could involve the investment in blood substitutes and artificial blood research, which may eventually reduce the dependence on human donors.

In the end, the ethics of blood donation hinge upon a delicate balance of compassion, respect, and scientific advancement. As we forge ahead, let us remember that at the heart of this issue beats a universal truth: our blood ties go beyond race, uniting us in the shared endeavour of preserving life. Through careful consideration and a commitment to ethical practices, we can ensure that this life-giving act remains a testament to the best of humanity.

Blood Politics: A Global Issue

In the shadow of these towering advancements, a silent crisis brews—a crisis of blood politics that shapes the fate of nations and the lives of millions. At the heart of this global issue lies an unsettling paradox: while some countries push the frontiers of blood science, others grapple with crippling shortages and an inequitable distribution of this vital resource. The stark reality is that access to blood and its products is a matter of life and death, yet it is governed by an intricate web of geopolitical forces and policies that often prioritise interests over human needs.

The visceral impact of blood shortages is felt across continents, from the bustling cities in the developed world to the most remote villages in developing nations. It's a grim tableau; emergency surgeries delayed, cancer treatments put on hold, and preventable deaths rising—all because the well of life runs dry. These shortages not only reflect a failure in health systems but also expose the deep-seated inequalities that define our global landscape. How can it be that in a world where we've mapped the human genome and engineered synthetic blood, countless individuals still face the spectre of death for want of a pint of blood?

Consider the story of Amina, a young mother in a rural African village, whose childbirth complications required an urgent blood transfusion. The nearest hospital, miles away, had exhausted its supply. Amina's plight is not unique; it's echoed in the lives of countless others, a personal struggle that lays bare the systemic shortcomings of the world's blood supply. Her story is a poignant reminder that behind the statistics and the policy discussions are real human beings, whose lives hang in the balance.

The stakes couldn't be higher. Every drop of blood is precious, every policy decision a matter of life and death. Yet, the international response to blood shortages and the management of blood supply is a tapestry of inconsistency and neglect. Blood, it seems, is not just a health commodity; it's a geopolitical pawn, subject to the push and pull of national interests, economic constraints, and international relations.

But there is a path forward, a way to navigate this complex landscape and emerge with solutions that transcend borders and politics. This book will unravel the tangled threads of the global blood crisis, examining the policies that dictate blood's flow and the innovations that promise to expand its availability. It will shed light on the efforts of organisations fighting to level the playing field and the role of diplomacy in fostering international cooperation on blood issues.

With unflinching clarity, we will dissect the economic models that determine the availability of blood, the role of private companies in the blood industry, and the impact of trade agreements on blood product exchange. We will explore the ethical considerations of blood donation and delve into the controversies surrounding paid versus voluntary donations, illuminating the cultural and societal factors that shape these practices.

As we peel back the layers of this issue, you will find that blood is more than a medical necessity—it's a mirror reflecting the health of our societies and the values we uphold. The blood crisis challenges us to rethink the way we share our most vital resources and to redefine what it means to be part of a global community.

Now, let us embark on this critical journey. Let us confront the stark realities and embrace the innovative solutions. Let us be spurred into action by the stories of those like Amina and let us rise to meet the high stakes with the gravity and resolve they demand. The path ahead is fraught with challenges, but it is also lined with hope—the hope that through understanding and cooperation, we can secure a future where blood ties extend beyond race, beyond borders, to bind us all in a shared commitment to preserving life.

The journey ahead will not be easy. It demands of us a willingness to challenge the status quo, to question long-held assumptions, and to engage in difficult dialogues. Yet, it is a journey that we must undertake if we are to honour the essence of our shared humanity.

The path ahead, however, is not without hope. This book is poised to embark on an exploration of solutions that can rectify the ethical imbalances inherent in blood banking. From policy reforms to technological innovations, from grassroots advocacy to international collaboration, the potential for positive change is tangible.

As the chapters unfold, tales of courage and innovation will emerge, stories that bear witness to the indomitable human spirit that drives us to improve, to evolve, and to care for one another. These narratives will serve as beacons of hope, illuminating the path towards a more ethical and equitable blood banking system.

Within these pages, you will find not just a critique of the present but a blueprint for the future—a vision of a world where blood ties extend beyond race, where every drop of blood is treasured, and where every life is given the chance to thrive. Join us on this quest for a moral renaissance in blood banking, a quest that promises to redefine what it means to give, to receive, and to share the very essence of life itself.

As you turn the pages of "Blood Politics: A Global Issue," prepare to be transported into the heart of a global conundrum, one that requires not just the keen insight of policymakers and the ingenuity of scientists, but the compassion and activism of every individual who believes in a world united by the common thread of humanity.

Together, let us forge a new understanding of blood's role in our world and take a stand in the face of a crisis that knows no boundaries. Let us write a new chapter in the story of blood—a chapter characterised by equity, innovation, and the indomitable spirit of global solidarity.

Blood on the Line: Medical Research and Morality

The essence of life flows through our veins, a crimson tide carrying the potential to heal, to save, to create miracles in the form of medical advancements. Yet, when the focus shifts to research laboratories, where blood is not just a gift but a commodity for scientific exploration, questions of morality begin to surface like ripples disturbing still water. What is the toll of progress, and at what point does the quest for knowledge cross the line of ethical integrity?

In the chambers of medical research, where the heartbeats of countless volunteers echo, we find our central theme: the utilisation of human blood for the sake of medical advancement. This narrative is not just about the blood itself, but the lives it touches, the boundaries it crosses, and the ethical spectre that looms over each vial collected.

The claim herein is bold yet necessary: while the use of blood in medical research has yielded extraordinary benefits, it is imperative that we maintain an unwavering commitment to ethical standards. We must navigate the precarious balance between innovation and morality, ensuring that each drop of blood serves the greater good without compromising the dignity and rights of the individuals who provide it.

To ground this claim, let us wade into the primary evidence. Medical breakthroughs such as the development of vaccines, the understanding of blood-borne diseases, and the advancement of surgical techniques are all indebted to blood research. These triumphs are concrete, their impacts measurable. Lives have been saved, lifespans extended, and the quality of life improved – these are the undeniable victories written in the language of blood.

Yet, to delve deeper is to uncover the layers of complexity shrouding this evidence. The stories are numerous: clinical trials where informed consent is murky, instances where economically disadvantaged individuals are disproportionately represented in blood donation for research, and situations where the risks may not be fully disclosed. Each narrative threads a needle of discomfort, prompting us to question the ethical frameworks governing medical research.

The counterevidence is stark, a sobering reflection of the times when ethical lines have been blurred. There are episodes in history where blood research has been conducted without proper consent or with coercion, where certain populations have been exploited, their blood drawn with promises unkept, or risks understated. It is an uncomfortable acknowledgment that progress has sometimes been achieved on the backs of the vulnerable.

Yet, in the face of these challenges, we must not lose sight of the capacity for reform. The rebuttal to these concerns lies in the strengthening of ethical oversight, in the rigorous application of informed consent, and in the equitable selection of research participants. These steps are not just remedial but foundational, ensuring that the integrity of blood research is upheld.

Additional supporting evidence emerges from the institutions that have embraced ethical reforms. Universities, hospitals, and research facilities have developed stringent ethical guidelines, review boards, and consent processes. They stand as testaments to the possibility of conducting blood research within the bounds of morality.

As we draw this discussion to a close, let us reinforce the assertion with unwavering conviction: the use of blood in medical research is a vessel of immense potential, but it must sail on ethical waters. The balance between the benefits of research and the ethical considerations is delicate, yet attainable. With vigilance, transparency, and respect for the sanctity of each individual's blood, we can ensure that the march of progress never tramples over the rights of the contributors to this invaluable resource.

In this quest, we are summoned to be both witnesses and guardians. We must bear witness to the stories of those who have given of themselves, often without recognition, and we must guard the future of blood research with an ethical compass that never falters. As the pages of this book turn, let them be a reminder of the blood ties that bind us, ties that go beyond race, beyond status, and beyond geography – ties that remind us of our shared humanity and our collective responsibility to honour it.

Personal Autonomy vs. Collective Benefit

In the tapestry of human society, where individual threads are interwoven into a complex pattern, the dichotomy between personal autonomy and the collective benefit is starkly revealed in the realm of blood donations. This delicate balance treads the line between the sanctity of individual choice and the urgent, often life-saving needs of the broader community.

Imagine, if you will, a world where the rivers of generosity run dry, where the wellsprings of compassion are capped by the steely resolve of personal prerogative. What becomes of the thousands, nay, millions who stand at the precipice of mortality, their lives hinging on the altruism of strangers? This is the crux of our current conundrum: ensuring a steady flow of voluntary blood donations while honouring the cherished principle of self-determination.

The marrow of the problem lies in the inherent tension between an individual's right to govern their own body and the societal imperative to aid those in distress. Blood, that vital elixir, is simultaneously a private good and a public necessity. The quandary we face is not merely theoretical; it is a practical challenge that demands our immediate attention and action.

Left unaddressed, the consequences of this standoff could be dire. Hospitals may find their reserves perilously low, surgery delays could become commonplace, and the spectre of blood shortages during emergencies might loom large. A societal failure to act could spell catastrophe for public health systems and, more crucially, for the individuals whose lives hang in the balance.

The solution, therefore, must be as multifaceted as the issue itself. It begins with an appeal to the core of what makes us human—the innate desire to help one another. Educational campaigns can illuminate the profound impact that blood donations have, not as abstract statistics but as narratives of survival and hope. By connecting donors to the faces and stories of those they help, the act of giving becomes personal, a shared journey rather than a solitary sacrifice.

Beyond the emotive, the practical steps to implementation are clear. Incentive programs, though contentious, could tip the scales in favour of donation. From small tokens of appreciation to tax benefits, these tangible rewards can serve as a catalyst for action. Moreover, removing barriers to donation—through mobile blood drives, extended hours, and simplified processes—can convert willingness into action.

Evidence of this solution's efficacy is not hard to find. Take, for instance, the blood drives that sweep through workplaces and communities, leaving in their wake reservoirs of life-saving blood. Consider the testimonials of those who, once recipients, now stand as donors, their gratitude manifesting as generosity. These are the real-world outcomes that underscore the viability of a strategy rooted in empathy and facilitated by convenience.

Yet, let us not be myopic in our approach. Alternative solutions merit consideration, such as harnessing the power of technology to create a more efficient blood supply chain or exploring synthetic alternatives to human blood. Such innovations could one day alleviate the pressure on blood banks and donors alike.

But let us return to the heart of the matter—to the pulsing question that runs beneath the surface of this discourse. What are you willing to give to save a life? Can you see the faces of those you might never meet, yet whose lives could be forever altered by your decision?

The march towards a harmonious resolution between personal autonomy and collective benefit is a journey that we must embark on together. It requires a chorus of voices, each singing a melody of compassion, each contributing to a symphony of solutions. It is a call to action that echoes in the silence between heartbeats—a call to which we must respond with open arms and open veins.

For in this delicate dance between the individual and the whole, every step forward is a step toward saving lives. As the pages of this discourse turn, let them inspire action, let them ignite a flame of collective responsibility. Let them be a testament to the power of unity in the face of adversity, a reminder that when we give a part of ourselves, we are not diminished but multiplied. Together, let us forge a path where the ties of blood become the bonds of a community united in purpose and in spirit.

Blood and Legal Identity

In the hallowed corridors of the High Court, the air was thick with anticipation. The undulating hum of hushed conversations filled the space, as barristers in their traditional garb navigated the maze of legal minds and anxious clients. This was a setting where the letter of the law met the untidiness of human life, and today, it was the crucible for a case that would intertwine the irrefutable evidence of blood with the complex web of legal identity.

At the heart of this legal theatre was a paternity dispute that had captured the public's attention. The main players: a young child caught in the crossfire of adult complexities, a mother seeking validation and support, and a man whose paternal link was questioned. The stakes were personal, yet the implications resonated far beyond the lives entangled in this drama.

The challenge laid bare before the court was one of biological truth versus legal standing. Was the man, indeed, the father of the child? Conventional wisdom and societal norms had been upended by the silent testimony of genetics, and now, blood would speak louder than words.

The approach taken to resolve this conundrum was a meticulous paternity testing procedure, a ballet of science danced in the laboratory where DNA was both witness and evidence. The strategy was simple yet profound: extract the genetic blueprint from each individual and let the sequences reveal the hidden bonds of kinship.

As the results were showcased, the impact was immediate and far-reaching. The genetic markers were conclusive, irrevocably altering the lives involved. A 99.9% probability of paternity was announced, echoing through the courtroom like the final gavel's blow. Data had spoken, and in its wake, the legal identity of a father was forged.

In analysing this case, one could not help but reflect on the broader implications. It highlighted the undeniable power of blood in determining identity and the way in which our legal frameworks adapt to the revelations of science. Yet, it also raised questions about privacy, consent, and the emotional toll of such irrevocable truths.

Connecting this single story to the larger narrative of 'Blood Ties: Beyond Race', it was clear that blood serves as more than a mere determinant of ethnicity or lineage. It is a powerful arbiter of identity, legal relationships, and, by extension, the rights and duties that flow from those bonds.

As readers turn the page on this chapter, a lingering thought remains: If blood is the key to unlocking identity, where do we draw the line between scientific certainty and the sanctity of human connection? The answer, much like the interplay of genetics and law, is bound in a complex dance of ethics, emotions, and the unyielding march of technological progress.

In the quiet aftermath of the court's decision, the young child, mother, and newly affirmed father faced a new reality. The blood that coursed through their veins had become the ink that would rewrite their stories, for better or worse. And as society grapples with these evolving truths, we must ask ourselves — are we prepared for the consequences that come when blood ties move beyond the realm of race, into the very heart of what it means to be legally bound?

The Business of Blood

Within the sterile walls of towering hospitals and the bustling clinics of urban streets lies an unseen lifeline, a commodity so vital yet so commonplace that its very ubiquity belies its significance. This is the world of blood: a market that operates silently alongside the more noticeable facets of healthcare, a realm where the essence of life flows not only through veins but through an intricate network of economic channels.

The issue at hand is pervasive and multifaceted, woven into the fabric of our healthcare systems. It concerns the blood industry, a sector that, though indispensable, is shrouded in complexities and ethical dilemmas. This industry is the lifeblood, quite literally, of medical practices, yet it sits at the intersection of altruism and economics, raising questions about the morality of profit derived from human biology.

Imagine, if you will, a world where this vital resource is diminished, where the blood banks run dry, and the supply cannot meet the demand. What would become of the surgeries put on hold, the treatments for cancer patients deferred, the lifeline for accident victims severed? The consequences are dire, painting a grim picture of a healthcare system teetering on the brink of a preventable crisis.

But let us not dwell on the precipice of disaster without casting our gaze towards the horizon of hope. For within this challenge lies the seed of opportunity, the chance to reform and to innovate. How, then, can we address this issue, ensuring a stable blood supply while navigating the moral quandary of commodifying that which courses through our own veins?

The answer, perhaps, lies in a multifaceted approach, one that honours the spirit of donation while recognising the need for a sustainable model. Consider the concept of value-based procurement, a methodology that places patient outcomes at the forefront of purchasing decisions. By adopting this model, healthcare providers could incentivise quality and efficiency among suppliers, ensuring that the business of blood, while still a business, serves the greater good.

Implementation of such a strategy would require a concerted effort, a collaboration between hospitals, blood banks, and policymakers. It would begin with a comprehensive review of current practices, followed by the establishment of clear guidelines that prioritise safety, sustainability, and ethical considerations. From there, the transition to value-based procurement could be rolled out gradually, with ongoing assessment and adjustment ensuring that patient care remains uncompromised.

But what evidence do we have that such a solution would be effective? Look to the handful of healthcare systems that have already begun this transition, where preliminary results show not only cost savings but improved patient outcomes. These early adopters serve as beacons, guiding the way towards a more ethical and sustainable blood industry.

Yet, it would be remiss not to acknowledge the other paths that lay before us. Alternative solutions abound, from advances in synthetic blood research to global initiatives aimed at increasing voluntary donations. Each option carries its own set of challenges and merits, and perhaps the true answer lies in a combination of approaches.

As readers, you might wonder, what role do you play in this narrative? Your actions, whether as a donor or an advocate for reform, can influence the trajectory of this industry. It is a reminder that the business of blood is not solely a matter for those in white coats and suits—it is a matter that touches all of us.

With each heartbeat, we are reminded of the pulse of life that connects us. In the business of blood, it is not merely a commodity to be traded but a sacred trust to be upheld. As we navigate the complexities of this industry, let us not forget the human element, the bonds that tie us together beyond the sterile transactions of the marketplace.

The story of 'Blood Ties: Beyond Race' continues, as we turn the page to yet another chapter, one that challenges us to confront the stark realities of our healthcare systems. It beckons us to ask ourselves: How can we ensure that the blood that binds us in our shared humanity is honoured not just in word, but indeed?

In the quiet reflection of these pages, a simple truth resonates: blood, in all its manifestations, ties us together in ways that transcend commerce. It is our collective responsibility to ensure that this bond remains unbroken, that the business of blood, while necessary, is never divorced from the sanctity of human life.

CHAPTER

5

The Myth of Race and the Science of Blood

In the annals of human history, the concept of race has carved deep chasms between cultures and continents. The genesis of these divides often lies in the shadows of antiquity, where empires and kingdoms were built upon the belief that lineage defines destiny. It is within these historical confines that the journey of debunking racial myths begins.

Picture a world not so long removed from our own, where the hue of one's skin dictated the scope of one's life. This was a time when pseudo-sciences such as phrenology claimed to discern intellect and morality from the shape of a skull—a time when eugenics sought to engineer a 'superior' race, leaving an indelible scar upon humanity's conscience. These were the fertile grounds where the seeds of racial hierarchy took root, intertwining with the fabric of societies across the globe.

The transatlantic slave trade, the colonisation of continents, and the subjugation of peoples; these were the milestones marking humanity's misguided quest for dominance. Nations rose to power on the backs of those they deemed inferior, crafting a narrative of racial supremacy that would echo through the ages.

Yet the march of time brings with it the relentless tide of change. Wars have been fought, laws enacted, and civil rights movements have risen, all striving to correct the course of history. From the abolition of slavery to the dismantling of apartheid, each victory has been a step toward the recognition of our shared humanity.

Fast forward to today, and the echoes of these historical battles still reverberate in our societies. Inequities persist, stereotypes linger, and the ghosts of bygone eras haunt the halls of progress. The need to understand the past is more than an academic pursuit; it is the key to unlocking a future where equality prevails.

Why, then, does history matter now? Because the past is the lens through which we can scrutinise the present and envision a better tomorrow. By understanding how the myths of racial difference were constructed, we can deconstruct the walls that these myths have maintained.

The story of human evolution is not one of separate lines but of a tangled web of migration, interbreeding, and shared survival. Our blood, a testament to this intertwined history, flows indistinguishable in its life-sustaining mission. Yet, even in the 21st century, the spectre of racial classification looms large, influencing everything from social interaction to medical care.

Could it be that the power of these antiquated ideas lies in their simplicity? In a complex world, the human mind seeks patterns and categories. Race, with its visible cues, has been an easy—but flawed—way to organise human diversity.

But let us not be misled by simplicity. The story of blood types, with its universal donors and its rare, accepting ABs, paints a picture of human connectedness that defies racial boundaries. It whispers a truth that is both simple and profound: we are more alike than we are different.

Imagine a society where the knowledge of our shared blood becomes the cornerstone of unity, a common ground upon which to build a more inclusive world. This vision is not a utopian dream, but a possibility grounded in scientific reality.

With each chapter, this book aims to peel back the layers of misapprehension and shine a light on the truth. It is a call to arms, not for conflict but for understanding, a plea for the recognition of our mutual bonds.

As the narrative unfolds, let it be a beacon guiding us out of the darkness of ignorance and into the light of knowledge. For in the end, it is not race but our shared blood that holds the key to our collective future—a future where the ties of blood movies beyond the realm of race, and where the myths of racial difference are finally laid to rest.

A Tale of cells and Societies

Blood types, those meticulous markers etched into our very biology, offer a profound commentary on human diversity. They are the unsung heroes in the narrative of life, more intimate to our being than the societal labels that often divide us. In the pages that follow, we delve into the rich interplay between the biological science of blood types and the societal constructs of race, two subjects that have shaped the course of human history and continue to influence the present.

Why compare blood types to race, one might ask? The purpose is to illuminate the stark contrasts and surprising parallels between an element as inherent as blood and a concept as abstract as race. We seek to understand how the former is an intrinsic truth of our existence, while the latter is a construct that has been imposed upon us, often with profound consequences.

The criteria for our comparison are clear. We shall consider history and impact, science, and society, the personal and the collective. We shall look at how both blood types and race have been used to unite and to separate, to save lives and to justify inequality. Our analysis will be meticulous, our exploration fearless, and our aim true: to shed light on the underlying threads that connect us all.

Now, let us directly compare. Both blood types and race are inherited. They speak of lineage and ancestry, of the silent legacy passed from one generation to the next. Just as blood types determine the compatibility for transfusions and organ donations, racial constructs have historically determined social status and privilege. Both can be viewed as methods of categorisation, creating groups based on certain characteristics.

Yet, when we direct our gaze to the contrasts, the distinctions become glaring. Blood types, though varied, do not predetermine one's abilities, character, or worth. They are free of prejudice. Race, on the other hand, has been burdened with stereotypes and biases, unjustly influencing perceptions and opportunities. Blood types unite us under the banner of biology, while racial constructs have often divided us under the guise of superficial differences.

Consider the visual aids of blood transfusion charts, showing which types are compatible, a clear and logical system based on science. There is no such chart for race because, scientifically, the concept of race holds no genetic or biological marker as definitive as blood types. Race is a mosaic, not easily delineated, its boundaries blurred by the brushstrokes of social, political, and historical context.

Our analysis reveals a profound insight: the way we perceive and act upon the concepts of blood type and race has far-reaching implications. Blood type compatibility is a matter of life and death, a scientific reality that can be tested and verified. Comparatively, race is a social construct, often wielded to justify inequalities, yet its origins and definitions are mired in ambiguity and controversy.

In contemporary relevance, the conversation is ever evolving. While scientific advancements in blood typing have refined medical practices, the dialogue on race is still fraught with tension and misunderstanding. In a world that is grappling with the legacy of racial discrimination, the objective clarity of blood types serves as a reminder of our shared humanity.

Have you ever stopped to consider the power of a drop of blood? How it can speak volumes about who we are, yet say nothing about the content of our character? Have you pondered the paradox of race, how something so visibly pronounced can be scientifically insubstantial?

Blood types do not shout their presence. They do not dictate our friendships, our loves, or the paths we walk in life. Race, conversely, can be a loud presence in society, influencing narratives and shaping destinies with a weight that belies its scientific ambiguity.

In the simplicity of blood types, we find a clarity that race, with all its complexities, cannot claim. As we navigate the intricate dance of cells and societies, let us remember the one-line truth that resonates through the ages: we are all human, bound by the same biological rhythms, regardless of the labels society has attached to us.

We must strive to embrace the simplicity in our biology and the complexity in our diversity. Let us use strong verbs like 'unite,' 'heal,' and 'understand' rather than 'divide,' 'harm,' and 'misinterpret.' Let our language reflect our intentions, and let our actions speak louder than words.

In the end, the tale of cells and societies is a continuous one. It is a story of discovery, of understanding, and, most importantly, of connection. As you close this chapter, may you carry with you the realisation that the essence of life flows within every one of us, a force that transcends the artificial boundaries of race, a force that reveals we are far more similar than we are different.

DNA Versus Melanin: What Really Matters

In the grand scheme of existence, what truly delineates one human from another? As the sun casts its impartial light on the rich tapestry of humanity, it highlights an array of skin tones, yet it is the invisible strands of DNA that weave the profound narrative of our species. Within the labyrinth of genetics lies the key to our individuality and our commonality, a concept far surpassing the superficial palette of melanin that colours our surface.

Does the blood coursing through our veins—a cocktail of A's, B's, O's, and AB's—carry more significance in determining our potential than the melanin dictating the shade of our skin? This question beckons us to delve into the genetic factors influencing human traits and capabilities, inviting us to explore a world where DNA reigns supreme over the pigments of our flesh.

Consider, for a moment, the vast array of abilities and proclivities encoded within the double helix of our DNA. From the resilience against certain diseases to the predisposition for particular talents, our genetic makeup plays a pivotal role in sculpting our physical and mental landscapes. It is within this molecular microcosm that the essence of our being resides—not within the melanocytes that merely paint our outward appearance.

Venture deeper into the realm of genetics, and you will find concrete evidence that our blood types, enzymes, and genetic markers exert a tangible influence on our existence. Take, for example, the compelling research on the correlation between blood types and susceptibility to diseases. Studies have shown that individuals with certain blood types are more prone to specific health conditions, while others may carry a sort of genetic shield against them.

But let us not pause there. For as we delve deeper into these genetic revelations, we uncover that blood types can even influence our body's response to different diets and how our blood clots, factors that can have a profound impact on our lifestyle and longevity. This evidence supports the claim that our blood—silent and unassuming—holds secrets to our survival and well-being; a silent sentinel in the odyssey of life.

What, then, of the counterarguments? Some might assert that race, with its visible cues and cultural significance, is a more impactful factor in one's life experience. It is true that society has, for centuries, constructed elaborate narratives and systems around the concept of race, often to the detriment and discrimination of many. The social implications of race are undeniable, shaping experiences and opportunities in nearly every corner of society.

Yet, upon closer examination, this social construct of race begins to unravel, revealing its lack of biological foundation. Genomics have uncovered the truth that all humans are genetically identical, rendering the concept of race virtually meaningless in a genetic and biological context. This revelation serves as a rebuttal to the weight society has placed on racial distinctions, highlighting the need for a paradigm shift towards a deeper understanding of our genetic unity.

Moreover, further supporting evidence comes from the scientific community's exploration into human migration patterns, revealing a stunning mosaic of interconnectivity. Geneticists can trace lineages across continents, illustrating not segregation, but an intricate web of shared ancestry. This, too, underscores the notion that what truly binds us is not the colour of our skin but the shared codes within our cells.

As we reach the denouement of this exploration, let us reinforce the assertion with a conclusion that echoes through the corridors of science and humanity alike. The assertion that our blood types, our genetic markers, are the true architects of our identity and capabilities holds steadfast against the superficial and divisive construct of race.

Let us not forget that our shared genetic heritage is the anthem of our species, a melody that transcends the arbitrary boundaries of skin colour. Let us carry forth the knowledge that within each of us pulses the same life force, a force that is blind to the distinctions that society has illuminated with such vigour.

In the final analysis, we are all woven from the same genetic cloth, a tapestry rich with the threads of diversity yet unified in its fundamental composition. It is this understanding that should guide our perceptions, inform our actions, and lead us toward a future where we celebrate not the colour of our skin, but the content of our genomes. For in the depths of our DNA lies the truth of our equality, the proof that we are tied by blood beyond the realm of our race, and the promise of a humanity viewed through the lens of science rather than the shadows of melanin.

Setting the Scene: Blood Types in the Modern World

In an epoch where the globe is intricately connected, the conversation about identity and what fundamentally separates or unites us is more pertinent than ever. Among the many elements contributing to this dialogue, one particular aspect of human biology stands out—not the colour of our skin, not the languages we speak, but the very lifeblood that courses through our veins: blood types.

The plot thickens when we place blood types within a broader framework, revealing how they intersect with various aspects of modern life. Medical professionals rely on knowledge of blood types to safely manage blood transfusions and organ transplants, and researchers study their epidemiological patterns to understand disease susceptibility. For instance, certain blood types have been linked to a higher or lower risk for developing conditions such as heart disease, Covid-19, and gastric cancer. Later we'll delve deeper under a topic "Blood Types and Global Health: A Public Perspective."

The practical applications extend beyond medicine. Consider the culinary world, where the "blood type diet" suggests that people should eat certain foods based on their blood type. Though this diet has its proponents, it underscores the fascination and sometimes, the misconceptions around blood types and their influence on our daily lives.

One common misconception is that blood type determines personality traits, a notion popularised in some cultures, particularly in Japan. However, scientific evidence does not support this belief, and it's important to dispel such myths to prevent the spread of pseudoscience.

Why does this matter? The narrative around blood types serves as a mirror reflecting the complex interplay between science and society. It reveals how we often seek ways to categorise and define ourselves, sometimes clinging to ideas that offer a semblance of identity or belonging, even in the absence of empirical support.

Are you aware of your blood type? Perhaps it's a question you've seldom pondered, yet it's a detail that could be crucial in a medical emergency. It's this silent, hidden information that might one day save your life, or the life of someone you love.

The discourse on blood types also touches upon ethical considerations. With certain blood types being rarer than others, the need for a diverse blood donor pool is critical. It's a call to action for communities to come together, transcending race and ethnicity, to donate blood and save lives.

In this dance of life, where every step is orchestrated by the complex symphony of our biology, blood types play a critical role. They are the unsung heroes in the narrative of human survival, a tale not of division, but of unity and shared vulnerability.

As we navigate the intricate tapestry of our existence, we must remember that the significance of blood types extends far beyond the confines of individual identity. It is a global story, one that intertwines with the threads of public health, community solidarity, and the very essence of what it means to be human.

Let us then embrace this understanding, cherishing the invisible bonds that connect us all. In the grand narrative of human existence, blood do indeed tie us together—perhaps even more than we have ever realised. They are a testament to our shared biology, a reminder that the essence of life flows within each of us, transcending the superficial distinctions that too often divide us.

Blood in the Age of Technology

In an era where technology advances at a breakneck pace, the realm of medicine stands as one of the most profoundly affected territories. It is a place where science fiction becomes science fact, and nowhere is this more evident than in the field of hematology—the study of blood. The red fluid that has been the life force of myth and legend is now on the brink of an unprecedented revolution. We stand at the threshold of a new age, an age where blood transcends the boundaries of biology and race, an age where the very essence of life can be synthesised, manipulated, and perfected. Welcome to "Blood in the Age of Technology."

Our journey begins with an objective clear as the plasma in your veins: to navigate through the labyrinth of modern medical advancements related to blood, to understand synthetic blood, gene editing, and their implications for the future of humanity. It is a quest to demystify the science, to bring into focus the potential for a new dawn of medical miracles, and to ponder the ethical landscapes that unfold before us.

Before we embark upon this odyssey, we must inventory our toolkit of knowledge. A rudimentary understanding of biology is a prerequisite, as is an openness to explore the ethical dimensions of scientific intervention. Familiarity with DNA—the blueprint of life—along with a grasp of genetic disorders and the current state of transfusion medicine will serve as the cornerstones of our intellectual arsenal.

As we stand at the entrance to this realm, let us take a moment to survey the landscape. Our broad overview encompasses two monumental advancements: the creation of synthetic blood and the advent of CRISPR-Cas9 gene editing. Both represent leaps toward previously unimaginable heights, where the limitation of blood supply and genetic blood disorders might one day become concerns of the past.

We now delve deeper, peeling back the layers to reveal the intricate tapestry of progress. Synthetic blood, a substance designed to mimic the function of human blood, holds the promise of an unlimited supply for transfusion, free of disease and compatible with all patients. Imagine a world where shortages are obsolete, where emergency rooms and battlefields benefit from this life-saving elixir.

The process of creating synthetic blood is a marvel of biotechnology. Scientists harvest stem cells, coaxing them into becoming red blood cells, platelets, or plasma. These artificial components can then be combined to form a complete and functional blood substitute. But the process is fraught with challenges. The cells must be cultured in precise conditions, and the resultant blood must undergo rigorous testing to ensure safety and efficacy.

As we venture further, we encounter gene editing, a beacon of hope for those afflicted with genetic blood disorders like sickle cell anaemia and thalassemia. CRISPR-Cas9, a revolutionary tool, allows scientists to cut and paste DNA with unprecedented accuracy, potentially correcting mutations at their source. The implications are staggering—curing the incurable, eradicating the ineradicable.

However, this power comes with great responsibility. The possibility of off-target effects, where unintended areas of DNA are altered, presents a significant risk. Moreover, the ethical considerations of gene editing in humans—especially in germline cells that would affect future generations—raise profound questions. What are the long-term consequences of tampering with the very fabric of life? Who decides which traits are desirable and which are not?

As we navigate these waters, we must offer tips and warnings. In the pursuit of synthetic blood, it is paramount to consider the immune response and the risk of rejection. Similarly, with gene editing, the temptation to enhance human capabilities beyond therapeutic needs—creating 'designer babies'—is a siren song that must be resisted to prevent a genetic divide in society.

Testing and validation of these technologies are critical. Clinical trials for synthetic blood must demonstrate safety and effectiveness, mirroring the circulatory system's complex functions. For gene editing, rigorous testing on cellular and animal models precedes human trials, and even then, long-term monitoring is essential to understand the full spectrum of effects.

Should you encounter turbulence along this journey—be it scepticism toward synthetic blood or ethical dilemmas surrounding gene editing—troubleshooting is key. Engage with the scientific community, participate in dialogues about the societal implications, and stay informed about the latest research findings.

In conclusion, is not a mere chronicle of scientific advancements; it is an exploration of the human condition, an examination of our moral compass as we stand on the precipice of altering life itself. As you turn the final page of this chapter, reflect on the questions posed and the images conjured. Consider the implications of a world bound not by the blood we're born with, but by the blood we choose to create and modify. What does it mean, then, to be human in the age of technology? And how do we navigate the delicate balance between innovation and ethics, between the possible and the permissible?

The blood that flows through our veins may be the same that has nourished our ancestors, but the blood of the future will tell a different story—a story that we have the power to write. Let us proceed with caution, with hope, and with a reverence for the sanctity of life's crimson thread, as we step into a new era where blood ties extend beyond race, beyond kin, into the very essence of what it means to be alive.

Myth-Busting: Race vs. Blood Type Prevalence

Eighty percent. That's the staggering proportion of South Americans and Central Americans who carry the type O blood allele. This number is not just a statistic; it serves as a powerful testament to the complex tapestry of blood type distribution across different populations, a pattern which often defies the neat, oversimplified narratives of racial biology. The very notion that certain blood types are inexorably tied to race begins to unravel under the scrutiny of hard data.

Why, you might ask, does this matter? Because in a world intricately woven with genetic threads, understanding the distribution of blood types transcends mere academic curiosity. It is a matter of life and death, a critical component in the efficiency of healthcare systems, the efficacy of medical treatments, and the global strategies for emergency preparedness. Yet, the importance of blood types is often overshadowed by the pervasive focus on race.

But what if I told you that the majority of African Americans are predisposed to have type O blood, while Asian populations have a higher prevalence of types B and A? Does this surprise you? It should. It hints at unseen depths of our collective genetic journey, a journey that intertwines with migrations, mutations, and historical encounters.

How, then, does this affect you, the reader, sitting with this book in your hands? What if your understanding of your own blood could unlock new perspectives on identity, health, and society? These are the questions that "Blood Ties: Beyond Race" seeks to answer, challenging you to rethink preconceived notions and embrace a more nuanced view of human diversity.

Imagine a world where blood type, not race, becomes a primary consideration in medical treatments, where emergency responders prioritize blood compatibility over racial profiles, where blood donation campaigns are tailored to genetic realities rather than racial myths. This is the enlightening journey that "Blood Ties: Beyond Race" promises to its readers.

As a medical science student at Edith Cowan University, my fascination with the intricacies of human blood has led me down a path of discovery—a path I now invite you to walk with me. Together, we will uncover the myths, reveal the realities, and understand the profound impact blood type has on our lives.

Why should you care? Consider this: every two seconds, someone in the United States needs blood and one in three Australians will need blood in their lifetime, but only one in thirty donates blood. That's a staggering demand that can only be met through a robust understanding and efficient management of blood type distribution. The relevance of this book cannot be overstated. It's not just about academic knowledge—it's about saving lives.

Could it be that we've been looking at blood all wrong? What if I told you that the universal donor, the cherished O negative, is not so universal when it comes to racial demographics? This book will reveal the surprising truths behind these assumptions, truths that could reshape the way we think about blood donation and transfusion.

And so, the journey begins. A journey that will take us from the historical origins of blood type classification to the cutting-edge of genetic research, from the common misconceptions that pervade society to the actionable insights that could transform healthcare.

This is not just a book. It is a call to action, a blueprint for change, and a beacon of knowledge in a sea of misinformation. It is a story that needs to be told, a conversation that must be had, and a truth that demands recognition.

The question now is, are you ready to be part of this change, to contribute to a future where blood ties extend beyond the realm of race? Let's embark on this journey together and see where the evidence leads us. The first step is to turn the page.

CHAPTER

6

Dietary Dialogues: Nutrition and Blood

In the ever-evolving discourse on health and wellness, an intriguing concept has emerged, whispering of a dietary revolution tailored to the very lifeblood coursing through our veins. Imagine, if you will, a world where the food on your plate is not just a matter of personal preference or cultural tradition, but a deliberate choice informed by the intrinsic nature of your blood type.

At the heart of this notion lies the blood type diet, a theory suggesting that the key to optimal health, weight management, and disease prevention may be found in eating according to the specific characteristics of one's blood type. This concept, while not without its sceptics, posits that each blood type—O, A, B, or AB—has its own distinct dietary needs, rooted in the evolutionary narratives of our ancestors.

Why, one might ask, would the type of blood flowing through our veins influence our nutritional requirements? The answer, proponents argue, lies in the biochemical makeup of the antigens present on the surface of our red blood cells. These antigens, it is theorised, interact with food components in a manner that can either promote health or trigger adverse reactions.

Consider the hunter-gatherer, with blood type O, thriving on a diet rich in protein from meats and animal products. Fast forward through the annals of time to the agrarian societies, where type A blood emerges, aligning with a more plant-based diet characteristic of agricultural abundance. The nomadic tribes of the steppes, their blood type B, found sustenance in dairy products, while the enigmatic AB type, a modern amalgamation, supposedly benefits from a balanced fusion of A and B dietary patterns.

In a world brimming with fad diets and quick fixes, the idea of type-specific nutrition offers a bespoke approach to eating. But how does one navigate this gastronomic map drawn with the ink of our own biology? Let us delve into the practical applications of this theory, illustrating with examples to bring clarity to the concept.

Picture Tolasa, a type O individual, who, after years of battling unexplained fatigue and digestive discomfort, finds solace in a diet that minimises grains and maximises lean meats and vegetables. His transformation is not just anecdotal;

it is echoed by others who share his blood type and report similar improvements.

Across the table, Emily, with type A blood, embraces a vegetarian lifestyle, rich in organic fruits, vegetables, and whole grains. For her, the elimination of red meat and an emphasis on fresh, natural foods aligns with her body's predispositions, yielding a sense of vitality and well-being.

The perspectives on this dietary approach are as varied as the blood types themselves. Critics argue that the scientific evidence supporting blood type diets is thin, suggesting that the benefits some experience may be the result of healthier eating patterns rather than blood type compatibility. Meanwhile, supporters cling to the personal testimonies and the logic of evolutionary adaptation as a testament to its validity.

Data supporting the blood type diet are indeed sparse, and yet, there are those who swear by its efficacy, bolstered by improvements in their biomarkers and overall health. In this quest for dietary truth, it is essential to dissect the available evidence, to separate the wheat from the chaff, as it were, in understanding the impact of blood type on nutrition.

Complex terms such as 'lectins'—proteins found in food that can agglutinate blood cells—and 'polymorphism'—the genetic variation that results in different blood types—require demystification. Lectins, for example, are said to interact differently with each blood type, potentially leading to inflammation or other health issues when consumed in discordance with one's blood type.

In concluding this exploration of type-specific nutrition, the key takeaways seem to hinge on personal discovery and scientific inquiry. While the blood type diet presents a fascinating framework for individualised nutrition, its full potential and scientific grounding remain areas ripe for further investigation.

A solitary question then lingers in the air, as potent as the aroma of a well-cooked meal: Could the secret to our health truly lie within the silent symphony of our blood? As we continue on this journey of nutritional discovery, let us savory the knowledge that comes from both ancient wisdom and modern science, ever mindful of the unique tapestries of our own bodies.

The A, B, AB, and O of Eating Right

As we venture further into the intricate dance of diet and blood type, the promise of personalised nutrition tailored to our genetic blueprint beckons with alluring clarity. Herein lies an opportunity to not only understand but to harmonise with the ancient rhythms that govern our bodies. Let us embark on a journey detailed with empirical wisdom and practical guidance, exploring the recommended dietary guidelines for each blood type.

The stage is set with four key players: the robust O, the agrarian A, the nomadic B, and the enigmatic AB. Each possesses its own dietary chorus, a symphony of foods that resonate with the genetic melody of its blood type.

At the dawn of humanity, where the Type O blood reigned supreme, a diet rich in protein played centre stage.

For those whose veins pulse with this ancient blood, a diet echoing the hunter-gatherer era is proposed. Embracing lean meats, such as grass-fed beef and poultry, is thought to fuel the Type O's robust digestive prowess. Seafood, a treasure trove of iodine-rich nourishment, also plays a pivotal role, particularly for those seeking to maintain optimal thyroid function.

Yet, it is not all flesh and fish. A bounty of leafy greens and vegetables complement the Type O's plate, providing essential nutrients and enzymes that facilitate digestive harmony. However, one must tread carefully amongst the grains and legumes, as they may play the discordant note for this blood type, potentially leading to inflammation and digestive disturbances.

Anecdotes abound of Type O individuals who, upon shifting their diets, report a surge in energy, a stabilisation of weight, and a newfound equilibrium. While empirical data may still be on the hunt, these personal stories provide a compelling narrative for those seeking to align with their ancestral diet.

How does this translate to the daily ritual of eating? Envision a breakfast plate adorned with a spinach omelette, a lunch of grilled salmon over mixed greens, and a dinner of lean roast beef with a side of steamed broccoli. These are not mere meals; they are sustenance strategies in tune with the Type O's evolutionary heritage.

As agriculture flourished, so did the presence of Type A blood, with a dietary pattern that reflects the bountiful yield of the earth.

Those graced with Type A blood are said to flourish on a natural vegetarian diet. A cornucopia of fruits and vegetables forms the bedrock of their sustenance, with whole grains and legumes acting as the supporting cast. This harmonious blend provides a spectrum of antioxidants, fibbers, and phytonutrients that cater to the Type A's immune system and digestive tract.

Dairy and animal proteins, while staples for some, are often suggested to be the antagonists for Type A's. They may contribute to digestive sluggishness and metabolic concerns, thus are advised to be consumed sparingly, if at all.

While the scientific community may still be cultivating the evidence, individuals with Type A blood who adopt this plant-centred diet often sprout tales of improved health markers and a blossoming of vitality.

Imagine a day beginning with a fortifying bowl of steel-cut oats adorned with berries, a lunch of quinoa and roasted vegetable salad, and a dinner of lentil stew. Each meal is a deliberate choice, an edible echo of the Type A's agricultural legacy.

With a lineage tracing back to the steppes, where mobility was key, Type B blood emerged with dietary flexibility as its hallmark.

The Type B diet is unique in its celebration of dairy. Yogurt, kefir, and cheeses are not just tolerated but embraced, providing a rich source of calcium and beneficial probiotics.

A balance of meat and plant-based foods characterises the Type B's optimal diet. Chicken, a common protein, is surprisingly cast as a villain for Type Bs, while game meats and lamb take the limelight. The inclusion of eggs, green vegetables, and certain grains like oats and millet completes the nourishing mosaic.

Stories from the Type B camp often recount a journey of dietary discovery, where the integration of moderate dairy and targeted protein choices leads to an oasis of health improvements, from metabolic stabilisation to mental clarity.

Envision a day where breakfast is a creamy Greek yogurt with honey, lunch is a roasted lamb salad, and dinner features a medley of grilled vegetables with venison. Such meals are not just sustenance but a celebration of the Type B's genetic heritage.

The youngest of the blood types, AB, is a fusion of A and B, and its dietary recommendations are as complex and intriguing as its lineage.

For the AB blood type, a diet that marries the best of both the A and B worlds is advocated. A mix of seafood, tofu, dairy, and an array of produce offers a tapestry of nutrients that cater to AB's adaptable digestive system.

While red meat is suggested in moderation, due to its potential impact on stomach acidity and digestion, plant-based proteins often take the centre stage. The eclectic AB can enjoy a diversity of foods, with a mindful approach to combinations and portions.

Again, while the empirical evidence for the AB diet is still being sketched, the canvas is filled with personal accounts of AB individuals who find a dietary sweet spot that echoes their genetic complexity.

Picture a day where breakfast is a vibrant fruit smoothie with a scoop of protein powder, lunch is a seafood salad, and dinner is tofu stir-fry with a rainbow of vegetables. Each meal is a stroke of dietary artistry, reflecting the AB's unique genetic blend.

In the symphony of eating right for your blood type, the notes may vary, but the melody is clear: harmony is found in the convergence of our ancestral echoes and the mindful choices we make at the table. As we savory each bite, let us remember that the journey to health is as individual as the blood that dances through our veins.

Debunking Diet Myths: Blood Type Versus Fads

In the realm of nutrition, the cacophony of dietary advice often drowns out the whispers of wisdom passed down through the ages. The modern diet landscape is a testament to the human penchant for trends, with each new fad sweeping across the globe like a tempest, promising a revolution in health and well-being. Yet, amid this storm of conjecture, a serene truth persistently beckons: the significance of our individual blood types may hold the key to unlocking the dietary harmony our bodies crave.

What if the well-trodden path of one-size-fits-all diets is a misstep? What if our focus on race and ethnicity as determinants of our nutritional needs is but a distraction from a more profound truth? The primary issue at stake is not one of heritage or hue but of haemoglobin and the blood that courses through our veins.

The consequences of ignoring this biological compass are stark. Misguided by generic dietary regimes, we risk aggravating our bodies' intricate systems, inciting a cascade of health ailments ranging from inflammation and allergies to obesity and chronic diseases. The potential for harm is not merely individual but societal, as the burden of preventable illnesses weighs heavily on healthcare systems worldwide.

But what of a solution? The method lies in embracing a diet that aligns with one's blood type, a blueprint that has evolved over millennia. The implementation of this strategy is both simple and profound: it begins with a shift in perspective, followed by a commitment to understanding and applying the principles of blood type-specific nutrition.

Imagine a world where you enter a grocery store armed with the knowledge of your blood type. As you navigate the aisles, your choices are guided not by the latest diet craze but by an ancestral map traced in your very cells. You fill your cart with foods that sing in harmony with your biology, and in doing so, you embark on a path to optimal health.

The evidence, though still emerging, is promising. Studies suggest that individuals who adhere to a blood type-specific diet report improvements in digestive health, weight management, and overall vitality. These outcomes, though anecdotal in nature, offer a glimpse into the potential power of personalised nutrition.

Can other solutions be entertained? Indeed, the realm of dietary wisdom is vast, and there are other approaches worth exploring. Some advocate for a plant-based diet for all, citing environmental sustainability and health benefits. Others extol the virtues of a ketogenic diet, a regimen high in fats and low in carbohydrates, which has shown promise in weight loss and metabolic health. While these alternatives hold merit, they lack the personalised touch that blood type-specific recommendations provide.

But how does one transition from the abstract to the tangible, from theory to table? The journey begins with education. One must seek out resources, perhaps a dietitian or a nutrition guidebook, that elucidate the connection between blood type and diet. From there, it is a matter of tailoring one's meal plans, experimenting with recipes, and observing the effects on one's health and well-being.

Have you ever paused to ponder whether the foods you consume are truly in sync with your body's needs? How often do you find yourself swayed by the siren song of the latest dietary trend, only to end up adrift in a sea of confusion and disappointment? The path to dietary enlightenment does not lie in the ephemeral but in the eternal – the timeless wisdom of our blood.

In the pages ahead, we will delve deeper into the specifics: the foods to embrace and those to avoid for each blood type, the scientific research that underpins these recommendations, and the real-life stories of transformation that lend credence to this approach. We will explore the nuances of meal planning, the challenges of dining out, and the strategies for navigating social situations where our dietary choices may be questioned.

Consider this: When was the last time you felt truly nourished by your meals, not just sated but deeply satisfied on a cellular level? The answers may be closer than you think, etched in the lifeblood that flows within you.

As we continue this exploration, let us not be swayed by the allure of adjectives and adverbs that merely decorate the surface of truth. Instead, let us seek the solid ground of verbs and nouns that carry weight and substance. Let us craft a narrative of nutrition that resonates with the very essence of who we are, painting a picture of health that is as vivid and varied as the human tapestry itself.

And so, we must ask: Are you ready to embark on a culinary journey that transcends the fads and fashions, a journey that is as unique as your fingerprint and as personal as your blood type? The table is set, the evidence is at hand, and the choice, as always, is yours.

Food for Thought: Case Studies

Nestled in the sprawling suburbs where the hustle of city life gives way to the tranquillity of tree-lined streets, our story unfolds. It's here, within the confines of ordinary existence, where extraordinary revelations about health and diet surface. The crisp morning air carries a promise of change, a subtle but powerful shift in perspective waiting to be discovered.

Meet Sarah, a middle-aged librarian with a penchant for historical novels and a history of digestive woes. Her colleague, Dr. Jonas, a spry nutritionist, often found himself engaged in discussions with Sarah about her health struggles. The players in this narrative are ordinary yet remarkable in their quest for well-being—a quest that would lead them down an unconventional path.

The challenge was clear: Sarah suffered from chronic fatigue, unexplained aches, and a general malaise that left her listless. Traditional dietary advice had failed her, leaving her on a carousel of elimination diets that only spun her further into confusion. It seemed her body was at odds with her intentions, and the discord was palpable.

Enter the solution, proposed by Dr. Jonas: a diet tailored to Sarah's blood type. With a blend of scientific curiosity and traditional wisdom, they embarked on a journey to align Sarah's nutrition with her O-positive blood type. It was a strategy rooted in the premise that certain foods could either be medicine or poison, depending on one's blood type.

The strategy was meticulous. Sarah began to incorporate lean meats, eschewed wheat and most grains, and embraced vegetables like kale and broccoli that thrived in her system. She avoided the nightshade vegetables that had previously left her joints aching and substituted them with beneficial alternatives.

The results were not immediate, but they were undeniable. Weeks turned into months, and Sarah's vitality began to blossom like a flower in the nourishing sun. Her energy levels soared, the aches receded, and her digestive system found its long-lost rhythm. The transformation was supported by data; her inflammatory markers, previously elevated, now painted a picture of internal harmony.

Reflection on this case reveals a tapestry woven with threads of scepticism and hope. Sarah had faced criticisms from those who dismissed blood type diets as lacking rigorous scientific validation. Yet, the empirical evidence of her own body could not be ignored. Her experience stood as a testament to the power of personalised nutrition, even as the scientific community continued to debate its merits.

Visual aids, such as Sarah's food diary and a chart tracking her symptoms and improvements, provided a graphical representation of her journey. They served as a tangible reminder that sometimes; the proof lies in personal experience.

The broader narrative here connects to the ancient wisdom that underpins the philosophy of eating for one's blood type. It speaks to the potential of personalised nutrition in a world where the individual is often lost in the masses. Sarah's case is a single thread in a larger tapestry, suggesting that the key to optimal health may be found within us, coursing through our veins.

So, one might ponder, could the secret to dietary success lie within our own biology? How many others, like Sarah, might find solace and vitality in a diet designed specifically for them? It's a question that lingers, inviting contemplation and perhaps even action.

As the author of this narrative, I have witnessed the power of individualised health strategies. Through the lens of academia and the prism of personal experience, the evidence before us urges a deeper exploration of the connection between blood type and diet.

As we turn the page, let us carry with us the curiosity that ignited Sarah's journey. May we embrace the diversity of our needs and the unique blueprints that guide us toward well-being. And may we continue to question, to learn, and to grow in our understanding of what it means to truly nourish ourselves. For in the end, it is not merely food that we seek, but the sustenance of life itself.

Digesting the Evidence: Scientific Support and Scepticism

In the ever-evolving discourse on diet and health, a contentious hypothesis suggests that the secret to optimal nutrition lies not in universal guidelines but in the unique imprint of our blood types. Such is the premise that undergirds the blood type diet, a concept that has both ardent supporters and vocal sceptics. In this exploration of the diet's validity, we aim to dissect the layers of scientific support and scepticism, presenting a balanced panorama of what we know, what we question, and what remains to be uncovered.

Amidst the cacophony of dietary doctrines, one voice stands out with a claim both specific and bold: that our blood type is the compass by which we should navigate the complex seas of nutritional choices. The proposition is straightforward yet profound: the alignment of diet with blood type can lead to better health, reduced disease risk, and an overall enhancement of well-being.

To anchor this claim in science, proponents point to the research that suggests different blood types have evolved to digest certain types of food more efficiently. For instance, the theory posits that those with Type O blood, being descendants of hunter-gatherers, thrive on a protein-rich diet, while individuals with Type A blood, allegedly from agrarian societies, benefit more from plant-based nutrition.

Diving deeper, proponents highlight studies where blood type antigens, the molecules that determine blood type, interact with lectins, a type of protein found in foods. These interactions, they argue, can affect everything from gut health to inflammation in the body. It is a compelling narrative, suggesting that the blood that flows through our veins carries with it ancestral dietary codes waiting to be deciphered.

However, a robust scientific inquiry necessitates the examination of counterevidence. Critics of the blood type diet point to the lack of large-scale, peer-reviewed studies that conclusively support the diet's efficacy. They argue that the anecdotal success stories, such as Sarah's, cannot be extrapolated to the general population without rigorous testing and reproducible results. Moreover, they contend that the diet's proposed evolutionary explanations are overly simplistic and not supported by anthropological evidence.

In response to these challenges, advocates of the blood type diet offer further clarification. They contend that while large-scale studies are scarce, the diet's principles are based on immunology and digestive biochemistry. They point to smaller studies and clinical observations that hint at the diet's potential benefits, urging the scientific community to conduct more comprehensive research.

The debate is enriched with additional supporting evidence from various fields. Some immunologists and nutritionists argue that the blood type diet could influence the microbiome – the complex community of microorganisms living in our digestive tracts. They suggest that blood type-specific diets might promote a more favourable balance of these microbes, thus contributing to better health.

Concluding this scientific odyssey, we reinforce our assertion that the relationship between blood type and diet warrants further investigation. Though the evidence is a patchwork of promising threads and unexplored patterns, the potential implications for personalised nutrition are too significant to dismiss outright.

One must ask: if our blood type holds the key to individualised health optimisation, isn't the pursuit of this knowledge a worthy endeavour? Can we afford to overlook a possible avenue to better health simply because it challenges conventional paradigms? These questions hang in the air, mingling with the scent of possibility.

The journey of understanding our nutritional needs is long and winding, and while the blood type diet may not be the panacea for all, it could very well be a beacon for some. As we strive to illuminate the path of dietary enlightenment, let us remember that science is not a static entity but a living process, constantly evolving as we feed it with curiosity, scepticism, and the relentless pursuit of truth.

CHAPTER

7

Personalities Pulsating Through Veins

In the grand tapestry of human understanding, few threads have woven as intricate and controversial a pattern as the concept of blood type-based personality typing. With origins obscured by the mist of time, this idea has emerged, not from the wellspring of modern psychology, but from the fertile minds of those who sought to categorise humanity by the very lifeblood that courses through our veins.

How did the belief that our blood type can determine our personality traits seep into the collective consciousness? It began, as many scientific inquiries do, with a spark of curiosity and a series of observations.

The earliest origins of blood type theories in personality psychology can be traced back to Japan in the early 20th century. It was there, in 1927, that Takeji Furukawa, a professor at Tokyo Women's Teacher's School, first postulated a connection between blood types and personalities in a paper titled "The Study of Temperament Through Blood Type." This marked the inception point of the subject, setting the stage for a cultural phenomenon that would endure for decades.

As the years progressed, Furukawa's ideas gained traction. His initial observations suggested that individuals with Type A blood were cooperative and sensitive, while those with Type B blood were passionate and creative. People with Type O blood was thought to be curious and generous, and those with Type AB were rational and calm. These character sketches formed the canvas upon which further theories would be painted.

The timeline of blood type personality theory is dotted with significant milestones. In the 1930s and 1940s, despite the backdrop of war and societal upheaval, the fascination with blood type and personality persisted in Japan. By the 1970s, books and articles discussing blood type personalities began to proliferate, reaching a wider audience. This surge in popularity coincided with the rise of self-help and personal development literature, providing a perfect storm of interest and opportunity for the theory to embed itself in popular culture.

The theory's reach extended beyond the borders of Japan, with cultural and regional variations manifesting in different ways. In South Korea and Taiwan, for example, blood type personality typing became just as popular as in Japan, with people seeking partners based on blood type compatibility, much like horoscopes in Western cultures.

In the modern era, these ideas have evolved and adapted, morphing into a blend of pseudoscience and entertainment. Television shows, books, and even dating services capitalize on the public's continued fascination with the concept. Blood type horoscopes and personality descriptors are now commonplace, offered with a wink and a nod, as much a topic of playful conversation as serious belief.

But what of the challenges, the controversies, and the turning points that mark the blood type personality theory's path? Scientists and psychologists have often debunked these theories, criticising their lack of empirical evidence and scientific rigor. The theory has been accused of perpetuating stereotypes, and in its darkest moments, it has been misused as a tool for discrimination.

Yet, one cannot deny the persistent allure of the idea that our physical makeup holds the key to unlocking the mysteries of our personalities. Could a drop of blood really hold the essence of who we are? Or is the blood type personality theory nothing more than a reflection of our desire to find patterns in the chaos of human behaviour?

The debate continues, even as new generations discover the theory and playfully ask, "What's your blood type?" as if the answer might reveal some profound truth about their character. In the end, perhaps the most telling aspect of blood type personality theory is not the veracity of its claims, but rather what it reveals about our collective yearning for understanding and the lengths we will go to find connections with one another.

As the chapter closes on this exploration of blood type and personality, one cannot help but ponder: What other mysteries of the human condition lie just beyond the veil of our current understanding, waiting to be unveiled by the curious and the bold?

Type by Type: Personality Profiles

Blood courses through our veins, a river of life that not only sustains us but, according to some, hints at the very essence of our personalities. The notion that one's blood type might mirror the intricacies of their character is as fascinating as it is contentious. This exploration ventures deeper into the personality characteristics traditionally associated with each blood type, offering a more nuanced portrait of this esoteric typology.

In the forthcoming sections, we will meticulously unravel the tapestry of traits attributed to each blood group. This journey is not just about listing qualities but understanding the layers of human complexity that they represent. Let us embark on this voyage of discovery, where each blood type reveals a unique constellation of characteristics that collectively contribute to the enigma of personality.

Blood Type A: The Conscientious Collaborator

Blood Type B: The Passionate Maverick

Blood Type O: The Assertive Trailblaser

Blood Type AB: The Rational Enigma

Blood Type A: The Conscientious Collaborator

People with Type A blood are often seen as the embodiment of precision and order. They are touted for their meticulous nature, a trait that manifests in a penchant for organisation and a keen attention to detail. This blood group is believed to house individuals who are cooperative and highly adept at creating harmony in group settings.

Delving deeper, we find that Type A personalities are not merely systematic but are often driven by a profound sense of responsibility. They take commitments seriously and are reliable to a fault. This sense of duty is complemented by a deep-seated sensitivity—an emotional compass that guides their interactions with the world.

But what evidence do we have to support these claims? Anecdotal testimonies abound, with countless individuals self-identifying with these traits. Yet, the scientific community remains sceptical, urging caution against overgeneralisation and reminding us of the complexity of human behaviour.

In practical applications, understanding the Type A personality can be beneficial in workplace dynamics. Managers might leverage their natural inclination towards organisation by entrusting them with roles that require meticulous planning and execution.

Blood Type B: The Passionate Maverick

The lore surrounding Type B individuals paints them as the free spirits of the blood type world. They are often described as creative souls, unbound by convention and driven by a thirst for innovation. This blood group is thought to harbor those who are passionate, spontaneous, and adaptable.

As we explore these traits, we uncover a portrait of individuals who are not afraid to walk the road less travelled. Type B's are said to be resilient in the face of change, making them adept at navigating life's unpredictable twists and turns.

Testimonials from those who resonate with these characteristics provide subjective corroboration, while sceptics point to the absence of empirical data to substantiate these correlations.

In real-world scenarios, understanding the Type B personality can be advantageous when seeking out individuals for roles that demand adaptability and out-of-the-box thinking. Their propensity for creativity makes them valuable assets in fields that prize innovation.

Blood Type O: The Assertive Trailblaser

Individuals with Type O blood is often characterised as confident leaders, with an innate ability to take charge. They are heralded for their assertiveness and are seen as natural-born motivators, capable of inspiring those around them.

Beneath the surface of this assertive exterior, Type O's are thought to possess a generous spirit and a capacity for boundless energy. This combination of traits suggests a personality that is not only ambitious but also altruistic.

While there is a plethora of stories from Type O individuals who identify with these qualities, the scientific lens casts a shadow of doubt on the validity of such broad characterisations.

In terms of practical applications, the Type O personality's leadership tendencies make them well-suited for positions that require decisiveness and vision. Their energy and motivation can be the driving force behind team success.

Blood Type AB: The Rational Enigma

Lastly, we turn our gaze to those with Type AB blood, a group often enshrouded in mystery. They are typically seen as the rational thinkers, cool and composed under pressure. This blood type is purported to produce personalities that are analytical and objective, capable of navigating complex situations with ease.

Digging into the psyche of Type AB's, we uncover a dual nature—a blend of the sensitivity of Type A and the adaptability of Type B. This duality allows them to approach issues from multiple angles, making them adept problem solvers.

Personal accounts from Type AB individuals lend credence to these descriptions, though, as with the other types, hard scientific evidence remains elusive.

In the realm of practicality, the Type AB personality's analytical prowess can be particularly valuable in roles that require strategic planning and critical thinking. Their calm demeanour allows them to serve as stabilizing forces within teams.

Do these blood type personality profiles truly capture the essence of our characters, or are they simply reflections of our desire for self-knowledge? Could it be that we are more than the sum of our biological components, our personalities shaped by an intricate web of experiences and choices?

Let us ponder these questions, for in the quest to understand ourselves, every theory offers a lens through which we might catch a glimpse of truth. The journey of self-discovery is unending, and perhaps, in the end, it is the seeking that matters most.

Cultural Perspectives: Blood Types in Social Contexts

In the intricate mosaic of human culture, there exists a phenomenon as curious as it is revealing: the belief in a connection between blood types and personality traits. This practice, deeply rooted in various societies, transcends mere folklore; it has woven itself into the fabric of everyday life, influencing social interactions, career choices, and even romantic relationships.

Blood type-personality correlation is a notion that suggests individuals' temperaments and dispositions are influenced by their blood group. This concept, while not scientifically substantiated, holds significant sway in certain cultures, where it is embraced with a fervour akin to astrological signs in the West.

To understand the gravity of this belief, it is essential to recognise its key elements. The idea posits that each blood type—A, B, AB, and O—carries with it a distinct set of personality traits. For example, Type A is often associated with conscientiousness and attention to detail, while Type B is linked to creativity and spontaneity.

The roots of this concept can be traced back to Japan in the 1920s, with the work of Takeji Furukawa, a professor who proposed that blood types could be used to assess the temperaments of potential employees. Over time, this notion expanded beyond the workplace and seeped into the broader cultural consciousness, influencing everything from matchmaking to parenting styles.

Placing this idea within a broader framework, it becomes evident that this belief system is not an isolated cultural quirk but rather one of many ways humans attempt to categorise and understand personality. From the four humours of ancient medicine to the Myers-Briggs Type Indicator, the quest to classify human nature is a universal endeavour.

Real-world examples of the blood type phenomenon abound, particularly in East Asian countries like Japan and South Korea. In these societies, it is not uncommon for blood type to be a topic of conversation, a factor in matchmaking, or even a segment in TV shows and newspaper columns dedicated to horoscope-like blood type forecasts.

Despite its widespread acceptance, the concept is not without its detractors who point out common misconceptions. Critics argue that the traits associated with blood types are often vague and general enough to apply to many people, a phenomenon known as the Forer effect. They also caution against the dangers of blood type discrimination, a form of prejudice akin to racism or sexism, where individuals might be judged or stereotyped based on their blood group rather than their individual merits.

Are you, perhaps, a Type A, meticulous and reliable, or a Type B, innovative and free-spirited? Or maybe you find yourself a blend, an AB, who can navigate complex problems with ease? Could your success as a leader be attributed to being a Type O? These questions, while intriguing, remind us of the simplicity and, yet the complexity of human identity.

In the grand tapestry of global cultures, the fascination with blood types as a determinant of personality is a thread that highlights our shared human desire to find patterns and meaning in the chaos of existence. It is a testament to our relentless pursuit of self-understanding and, perhaps, a reminder of our tendency to seek simplicity in the intricate dance of nature and nurture.

While the scientific jury may still be out on the validity of blood type-personality correlations, the cultural impact is undeniable. It provides a unique lens through which to view ourselves and others, a framework for interpreting behaviours, and a narrative to which many have attached their identities. So, as we navigate the complex interplay of genetics and personality, let us do so with an open mind and a critical eye, ever mindful of the rich cultural narratives that colour our perception of the world.

The Science of Sanguine Traits: Fact or Fallacy?

Blood types have long held a place of fascination in the human imagination, a symbol of the invisible forces within us that may shape who we are. The quest for understanding has led some to posit that these markers of our biology are not mere agents of coagulation but keys to unlocking the mysteries of personality. Yet, the leap from physiological functions to psychological predictions is a chasm that requires not just a bridge of hypotheses but the solid ground of empirical evidence.

It is within this context that we scrutinise the claim that blood ties move beyond the realm of race in determining personality. This proposition is not only a significant issue but one that implies profound consequences for our social dynamics and personal self-conception. The crux of this contention lies in the assertion that one's sanguine classification might dictate their temperament, a notion that, if unfounded, could lead to a myriad of misconceptions and discriminatory practices.

Picture, if you will, a world where blood type dictates one's destiny. Employment opportunities, romantic engagements, and social circles might all hinge on an A, B, AB, or O. This scenario is not mere science fiction; it mirrors reality in some cultures where such beliefs hold sway. Yet, what implications might we face if these beliefs are embraced without scientific merit? The potential for blood type prejudice, akin to racial or gender discrimination, looms large—a spectre of division based on unvalidated science.

To confront this challenge, we must turn to rigorous scientific inquiry. The solution lies not in the reinforcement of unproven correlations but in the dissemination of robust, evidence-based information. This approach necessitates comprehensive research studies that examine the purported links between blood type and personality with a critical eye, employing large, diverse cohorts and stringent controls to rule out confirmation bias and the Forer effect.

Implementing this solution would require the mobilisation of interdisciplinary teams, from geneticists to psychologists, to conduct expansive and replicable studies. Peer-reviewed publications would then disseminate their findings, gradually replacing folklore with facts. Educational campaigns could further enlighten the public, dispelling myths with accessible summaries of the scientific consensus.

Evidence of the efficacy of this approach can be found in the history of other pseudoscientific beliefs. Consider, for example, the debunking of phrenology—the idea that one could determine personality from the bumps on the skull. Through scientific rigor, this notion was relegated to the annals of debunked theories, a fate that could befall the blood type-personality hypothesis should it fail to withstand empirical scrutiny.

Yet, what if there is a grain of truth to be found within the chaff of conjecture? Could there be alternative solutions or explanations that warrant consideration? It is crucial to evaluate all facets of the argument. For instance, might there be a psychosomatic element at play, where belief in the blood type-personality link influences behaviour, creating a self-fulfilling prophecy? Exploring such avenues could yield insights into the placebo effect and the power of belief on human psychology.

Imagine a world emancipated from the shackles of unsupported claims. Where scientific literacy prevails, and individuals are seen not as a collection of type-specific traits but as unique and multifaceted beings. In such a world, the content of one's character is not pre-determined by the blood coursing through their veins but by the myriad experiences and choices that comprise their life story.

Are you not, after all, more than a simple letter or combination thereof? Is it not within the rich canvas of your experiences that your true personality is painted? Let us then embark on a journey of discovery that honours the complexity of the human spirit, unbound by the simplistic notions that threaten to confine us to predetermined boxes.

In conclusion, as we delve deeper into the science of sanguine traits, we must do so with a commitment to truth and a resolve to prevent the propagation of fallacies. The narrative of blood type-personality correlation serves as a poignant example of humanity's enduring quest for self-definition but also as a cautionary tale of the perils of embracing appealing narratives without empirical support. Thus, we stand at the crossroads of belief and knowledge, tasked with forging a path that respects both the allure of mystery and the imperative of evidence.

Personality Case Studies: Blood Types in Action

In the vast and varied tapestry of human experience, it is the singular stories—the case studies of individual lives—that often resonate most deeply with our quest for understanding. So let us draw back the curtain on an intriguing tableau, one where the interplay of blood types and personality promises to reveal secrets as intimate as the very blood that flows in our veins.

Consider a bustling urban landscape, an environment where the hum of activity is as constant as the rhythm of beating hearts. In this dynamic setting, we encounter a unique array of individuals, each navigating the complexities of modern life with a blood type that silently accompanies them, an unseen companion on their journey.

Among this diverse cast, we find Maya, an ambitious professional with blood type O, and Hiroshi, a thoughtful artist who wears blood type A not on his sleeve, but deep within. Then there's Sofia, a spirited social worker of blood type B, and finally Eli, an AB-blood type theoretical physicist, whose mind dances among the stars while his feet tread on terra firm.

The heart of our exploration beats with a provocative question: Does the blood that courses through Maya, Hiroshi, Sofia, and Eli's veins impart distinct personality traits that ripple across the surface of their lives, influencing their interactions, aspirations, and even their conflicts?

To seek answers, we delve into their stories, examining how each individual confronts life's hurdles. Maya's O-type drive and leadership qualities see her rallying her team through a critical project deadline. Hiroshi's A-type attention to detail and sensitivity help him navigate the complex emotional landscape of a collaborative art installation. Sofia's B-type independence and adaptability shine as she orchestrates a community outreach program. And Eli's AB-type analytical prowess and creativity are put to the test in the pursuit of a ground-breaking scientific discovery.

The outcomes are as varied as their personalities. Maya's project is a resounding success, reinforcing the association of type O with assertiveness and resilience. Hiroshi's art exhibit receives acclaim for its emotional depth, echoing A-type introspective qualities. Sofia's program thrives, her B-type versatility proving invaluable. Eli's research breaks new ground, his AB-type analytical nature playing a pivotal role.

Yet, we must pause to consider: Are these successes truly a reflection of their blood types, or are they the product of individual experiences, skills, and choices? The evidence before us is anecdotal, compelling in narrative but not conclusive in science. We tread carefully on the path of correlation, wary of mistaking it for causation.

Blood Type A: The Conscientious Collaborator	Meticulous, organized, and detail oriented. Cooperative and skilled at creating harmony in group settings. Driven by a sense of responsibility and reliability. Sensitive and guided by an emotional compass
Blood Type B: The Passionate Maverick	Creative, innovative, and unbound by convention Passionate, spontaneous, and adaptable Resilient in the face of change and adept at navigating unpredictability Valuable assets in fields that prize innovation
Blood Type AB: The Rational Enigma	Rational thinkers who remain composed under pressure Analytical, objective, and skilled at navigating complex situations. Possess a dual nature, blending the sensitivity of Type A and the adaptability of Type B Valuable in roles that require strategic planning and critical thinking
Blood Type O: The Assertive Trailblazer	Confident leaders with a natural ability to take charge. Assertive and capable of inspiring others. Ambitious and altruistic. Well-suited for positions that require decisiveness and vision.

Graphs depicting personality traits distribution across blood types pepper our analysis, inviting the eye to question as well as see. Yet, these visual aids serve as reminders that correlation does not imply causation, and the complexities of human personality resist simplification.

These stories, these snapshots of life, serve not as definitive proof but as a mosaic of possibilities. They add colour and contour to the broader question of whether blood types can indeed shape the essence of who we are.

One might ponder, is it plausible that these blood-based archetypes have merit, or are they simply the modern echoes of ancient attempts to categorise the human soul?

As I pen these words from the quiet confines of Edith Cowan University, where the pursuit of medical science is steeped in scepticism and the thirst for empirical truth, I am reminded that the journey to understanding is never-ending. Each story, each case study, is but a piece of the puzzle, a fragment of the grand narrative that defines our shared humanity.

Could it be that the blood flowing within us whispers truths about our character, or is it merely the silent witness to the unfolding drama of our lives? The answer remains shrouded in the mist of uncertainty, yet the quest for clarity is a noble and necessary endeavour.

As we close this chapter of our exploration, let us not succumb to the allure of easy answers but continue to question, to probe, and to challenge the narratives that seek to define us. For it is in the crucible of inquiry that the ore of knowledge is refined, and from this process, a greater understanding of ourselves and each other is forged.

Thus, we are left to ponder, to reflect on the stories we have shared, and to ask ourselves: Are we the sum of predetermined types, or are we architects of our own identities, with blood type as but one thread in the rich and complex weave of our existence?

CHAPTER

8

The Heartbeat of Sexuality

Have you ever considered the mysterious alchemy that sparks romance? What if the secret to compatibility lies not in the stars, nor in the depths of our psyches, but within the very life force that courses through our veins—our blood?

Within the vast tapestry of human connection, the notion that blood types could influence sexual and romantic compatibility is a thread that's been largely overlooked. Yet, as we delve into the intricate dance of attraction and affection, this concept beckons with a strange allure, a siren calls to those yearning to decode the enigma of love.

Consider for a moment the central issue this poses: The quest for a compatible partner is as old as time itself, and myriad factors are touted as keys to unlocking this perennial puzzle. Personality traits, shared interests, and physical attraction are the usual suspects in this matching game. However, what if there's a biological undercurrent, a silent player in the game of love that we've yet to give its due?

The common misconception is that compatibility is solely a matter of heart and mind. Yet, here I present to you a unique perspective, one that stems from the lifeblood of scientific curiosity. As a medical science student, I've been taught to question, to research, and to look beyond the obvious. So, why not apply this approach to matters of the heart?

Think about it. When it comes to choosing a partner, most people rely on the tangible – the touch, the gaze, the conversation. But what if our blood type could whisper secrets of compatibility that our conscious minds barely grasp? The idea that lovers with certain blood type pairings might experience a more profound connection is not just intriguing; it's revolutionary.

Now, before you dismiss this as mere fantasy, consider the biological symphony that defines us. Blood type has been linked to personality traits, disease susceptibility, and even diet efficacy. It's not far-fetched to ponder its role in the chemistry of love. Imagine a world where a simple blood test could hint at the potential for passion and understanding between two individuals.

Let's forge a real connection here. Think of your past relationships, the ones that soared and those that sank. Can you completely attribute their success or failure to mutual interests or emotional rapport? Or might there be an invisible biological factor at play, one as fundamental as your blood type?

This isn't about reducing romance to a clinical procedure. It's about enriching our understanding of human connection by acknowledging the power of our physiological makeup. It's about recognising that perhaps our blood types can offer clues to the heart's enigmatic desires.

Picture this: Two individuals, seemingly perfect for each other, yet their relationship lacks that inexplicable spark. Could it be that their blood types are at odds, creating a subtle but insurmountable barrier to true intimacy? Conversely, consider a couple, inexplicably drawn to each other despite their differences, sharing a blood compatibility that ignites an undeniable fire.

To engage emotionally, let's not view this as a cold, scientific hypothesis. Instead, see it as a missing piece of the puzzle, a new dimension of compatibility that could lead to deeper, more harmonious unions. Imagine the possibility of relationships not just surviving but thriving, bolstered by this newfound understanding of our biological compatibility.

And so, I invite you to join me on this journey of discovery. Let's explore together the rich landscape where love and science intersect, where the beating of our hearts may well echo the compatibility of our blood. For in the grand tapestry of human connection, every thread has its place—and perhaps, blood ties us beyond the realm of our race—perhaps more than we've ever imagined.

Attraction in the Blood: Theories and Evidence

In the quest to decode the enigma of attraction, we turn our gaze to a scientific perspective that considers the very essence of our being—our blood. The proposition, at once controversial and compelling, posits that the compatibility of blood types plays a significant role in human attraction, perhaps even more so than the concept of race. But how much weight does this theory hold in the face of scientific scrutiny?

The claim that beckons our curiosity is this: blood types, determined by the presence or absence of certain antigens on the surface of red blood cells, have the potential to influence interpersonal chemistry and romantic compatibility. This assertion is not without its allure; after all, blood types, like fingerprints, are unique to each individual, and their influence on health and personality has been a subject of debate for decades.

In considering the primary evidence for this claim, we find a fascinating blend of cultural beliefs and scientific studies. In Japan, for example, the idea that blood type can dictate personality and compatibility is so ingrained that it's commonplace for individuals to mention their blood type in dating profiles, much like horoscope signs in the West. But does this cultural phenomenon have a basis in scientific fact?

To delve deeper, we must look at what research has to say. Studies have suggested that certain blood types may be more susceptible to particular diseases, and that blood antigens can influence the immune system in ways that might affect personality. A leap from these findings to the realm of attraction, however, requires a bridge of evidence that has yet to be fully constructed.

And yet, there are counterarguments to consider. Critics of the blood type attraction theory point out that there is a significant lack of empirical data supporting the idea that blood types can have a direct impact on romantic connections. They argue that attraction is too complex, a tapestry woven from the threads of social, psychological, and environmental factors that blood types alone cannot explain.

In response, proponents of the theory offer clarification, citing that while blood types may not be the sole determinant of attraction, they could play a role in the subtle interplay of biological compatibility. For example, it's known that blood type can affect pheromones, and since these chemical signals are crucial in human attraction, could there not be a link worth exploring?

Though the rebuttal holds promise, it is clear that additional supporting evidence is necessary to fortify the claim. Anecdotal accounts and cultural practices provide a foundation, but for a theory to stand, it must be supported by a framework of rigorous scientific validation.

In conclusion, the assertion that in the blood ties: beyond the realm of individual's race and ethnicity in matters of attraction is a fascinating one, offering a fresh lens through which to view the complexities of human relationships. The evidence currently available provides intriguing breadcrumbs on a path that is yet to be fully mapped. As we continue to explore the intricate dance of attraction, it is essential to remember that love is a multifaceted jewel, reflecting light from a myriad of angles. While blood types may indeed hold a piece of the puzzle, the picture is incomplete without considering the rich diversity of factors that shape the human heart. In this journey of discovery, let us remain open to the myriad possibilities that science and experience unfold before us, for in the end, the quest to understand love is as boundless as love itself.

Compatibility Quizzes: Blood Types in Love

In the soft amber glow of a Tokyo summer evening, two figures sat across from each other in a cosy izakaya, their knees brushing under the low table. Mika, a graphic designer with an eye for symmetry, had always been meticulous about the details of her life, from the way she organised her bookshelves to the precision with which she matched her socks. Across from her was Kenji, a soft-spoken biologist whose curiosity about the natural world was only matched by his passion for uncovering its mysteries.

Their conversation flowed as smoothly as the sake they sipped, touching on everything from their favourite films to their most cherished childhood memories. It was their first date, but the connection between them felt as familiar as an old song. The evening was warm, the mood was light, and yet, beneath the laughter, something profound was taking place. Mika and Kenji were unwittingly embarking on a journey that would challenge the very notions of compatibility they had always taken for granted.

As the night wore on, Mika's playfulness gave way to a moment of earnest curiosity. "So, what's your blood type?"

she asked. In Japan, this question was as common as inquiring about one's zodiac sign. Kenji replied with a shy smile, "I'm type O." Mika's eyes widened. "Me too!" she exclaimed. In that instant, a bond seemed to form, an invisible thread woven from the tapestry of shared biology.

The conversation that followed delved into the heart of what had brought them together. Was it merely coincidence, or was there something deeper at play? They shared a laugh over the popular personality assessments linked to blood types—how type O was said to be confident and self-determined, perhaps explaining their individualistic streaks. But as a medical science student, I've always been sceptical about such generalisations, preferring to lean on the tangible pillars of scientific evidence.

And yet, as I sat and made my observation about Mika and Kenji, I couldn't help but wonder: was there a kernel of truth to the lore that surrounded blood type compatibility? Could the same blood coursing through their veins be a silent signal, harmonising their attraction in a way that transcended cultural stereotypes?

This book, "Blood Ties: Beyond Race" is a journey to uncover that truth. Through its pages, I invite you to explore the mysterious link between blood types and relationships. We'll navigate through a labyrinth of cultural beliefs and scientific inquiries, examining how something as intrinsic as the blood we carry could influence the heart's desires.

Could the clumping of red blood cells under a microscope really dictate the clashing or meshing of two souls? Is there an undercurrent of truth to the belief that type A's sensitivity and type B's creativity could lead to a harmonious or tumultuous love affair? How much of this is rooted in biology, and how much is woven into the fabric of cultural mythology?

As we journey together, I'll share with you stories like that of Mika and Kenji, tales that resonate with the universal quest for understanding and connection. We'll ponder the subtle dance of genetics and attraction, seeking the wisdom hidden within the chemical ballet of our bodies.

"Can the essence of love really be distilled into a single drop of blood?" you might ask. Let's find out together.

In this exploration, I promise you insights gleaned from both the laboratory and the heart. We will dissect the claims, scrutinise the evidence, and, perhaps, arrive at revelations that will reshape our understanding of compatibility. After all, love is an enigma, and every discovery brings us closer to its core.

So, let us embark on this unexpected journey, where the only certainty is the excitement of the unknown. Where the questions we ask are as important as the answers we seek. Will we find that indeed blood ties beyond race in love? Only time, and the pages to come, will tell.

Blood Bound: Case Studies on Intimate Relationships

In the quaint town of Fremantle, Western Australia, a brisk ocean wind carries the laughter of children and the chatter of marketgoers. The scent of salt and freshly brewed coffee intermingles, crafting a tapestry of the local ambiance. At a quiet café, nestled between the vibrant stalls of the Fremantle

Markets, sat Julia, a nurse with a compassionate spirit, and Alex, a marine biologist whose life's work was to decode the secrets of the ocean's depths.

They were an unlikely pair, Julia with her pragmatic views on life and Alex with his head often in the clouds, dreaming of his next underwater adventure. Their backgrounds were as diverse as their personalities, Julia of Italian descent, and Alex, a native Australian with a lineage tracing back to the indigenous Nyoongar people.

The challenge that lay before them was unique. Julia, with her blood type AB, had always been fascinated by the concept of blood type compatibility, a topic she often encountered in her nursing career. And Alex, with his type A blood, had never given it much thought until he met Julia, who made him ponder the deeper connections between people.

Their approach to this challenge was methodical. Julia introduced Alex to the world of blood type personalities, explaining the traits often associated with ABs as diplomatic and rational, and those tied to as meticulous and responsible. Together, they began an exploration into how their blood types might influence their relationship dynamics.

The results were intriguing. They found that their conversations often took on a pattern that mirrored the supposed traits of their blood types. Julia would mediate and consider all sides of a discussion, while Alex would meticulously plan their weekend getaways with precision. They even noticed that their arguments rarely escalated, as their communication styles were complementary.

As I delved deeper into their story, I couldn't help but analyse and reflect upon the broader implications. Were these patterns a coincidence, or did their blood types genuinely play a role in their compatibility? It was clear that their relationship thrived on mutual respect and understanding, but how much of that was influenced by their biological makeup?

Connecting their story to the larger narrative of blood type compatibility, it became evident that while blood types might offer a framework for understanding personality traits, the human experience is far too complex to be solely defined by such parameters.

As I reflect on Julia and Alex's relationship, a thought lingers in my mind, prompting further contemplation. How do our biological blueprints shape the way we interact with those we hold dear? Is it possible that the fluid that gives us life also holds the keys to understanding our hearts?

As a medical science student, I am trained to seek answers through empirical evidence. Yet, the stories of couples like Julia and Alex remind me that the intricacies of human relationships often defy clear-cut explanations.

In this exploration, we must remain open to the possibility that the dance of love involves more than just the physical and the observable. It's a complex choreography where emotions, experiences, and perhaps, even blood types, lead us through the steps.

Could the compatibility of blood types truly influence the way we forge bonds, or is it merely one piece of the vast puzzle that is human connection? Is it possible that the same blood coursing through our veins sings a silent serenade to potential partners, or are we simply seeing patterns where none exist?

As we continue to unravel the enigma of intimacy and compatibility, we must ask ourselves: what truly binds us to one another? Is it the blood we share, or the countless moments that weave the rich tapestry of our shared lives?

Let us not forget that love, in all its forms, is a mystery to be experienced rather than solved. And as we turn the page to the next chapter, let's ponder the silent symphony of biology and emotion that plays within each of us, guiding us towards connection, understanding, and perhaps, even love.

Blood Type and Sexuality: Separating Myth from Science

In a world brimming with diversity, the human inclination to seek patterns and connections is ever-present. Our quest to understand the intricate tapestry of human interaction often leads us down paths lined with conjecture and folklore. Among the most enigmatic of these paths is the belief that blood type influences sexuality—a claim that has piqued curiosity and scepticism in equal measure.

Venture deeper into this narrative, and one may find oneself entangled in a web of myths masquerading as scientific fact. The notion that one's blood type can dictate sexual behaviour or preference is a topic fraught with controversy and misunderstanding. It's a realm where science and myth intertwine, often leaving the uninformed in a state of confusion.

Imagine, if you will, a society where blood types are heralded as the compass of compatibility, where the essence of attraction is distilled into the letters A, B, AB, and O. What consequences might arise from such a belief? Might we see discrimination based on this physiological trait, echoing the unjust biases of racism but with a new, pseudoscientific veneer?

Now, let's turn our gaze to a solution. To dispel these misconceptions, one must arm themselves with empirical evidence and critical thinking. It is through rigorous scientific inquiry that we can begin to dismantle the myths surrounding blood type and sexuality.

Consider the steps required to uproot such deep-seated beliefs. Education, undoubtedly, is the cornerstone. By integrating comprehensive sex education with a strong foundation in genetics and biology, we can enlighten minds and encourage scepticism toward baseless claims. Furthermore, promoting open discourse in public forums can aid in debunking myths and fostering a culture that values scientific literacy.

But what evidence do we have that these solutions could be effective? Look to the countless studies that have examined the link between blood type and personality, many of which have found no substantial evidence to support the claims. Consider the success of educational campaigns in other areas of public health that have shifted public perception and behaviour.

Could there be alternative solutions to address this issue? Certainly, one might consider leveraging technology to disseminate accurate information. Social media, apps, and websites could be powerful tools in spreading awareness and countering misinformation.

As we navigate this complex issue, let us pose a direct question: Are we, as a society, willing to let go of appealing yet unfounded beliefs in favour of hard evidence? To those who hold tightly to the romanticism of blood type dictating love and desire, this may seem a cold and unfeeling transition. Yet, it is within the realm of science that we often find the true beauty of our human experience—richer and more intricate than any myth could capture.

The exploration of blood type and sexuality serves as a mirror, reflecting our broader struggles with separating fact from fiction in the quest to understand ourselves and one another. It is a reminder that our individual experiences, shaped by a myriad of factors both biological and environmental, are too vast to be confined to simplistic categorisations.

In the end, the narrative of blood type and sexuality is not simply about debunking a myth. It is about embracing the complexity of human nature, celebrating our diversity, and recognising that the bonds we form are woven from far more than the markers in our veins.

The journey toward knowledge is ongoing, each step taken illuminating the path ahead. Let us proceed with curiosity and courage, unafraid to challenge the myths that obscure our understanding, and ever ready to embrace the truths that unite us in our shared humanity.

And so, we return to the essence of our inquiry, to the heart of the matter: the recognition that love and attraction are not dictated by blood, but by the myriad qualities that make us uniquely human. It is through this lens that we must view ourselves and others, not as types or categories, but as individuals whose capacity for connection transcends the simplistic and enters the realm of the profound.

Let this be the message that resonates, a clarion call for reason and compassion in a world too often divided by the superficial. May we all find the strength to see beyond the myths, to the core of what truly matters—the content of our character, the depth of our emotions, and the unquantifiable essence of our humanity.

CHAPTER

9

Motherhood and Maternity: The Bloodline Connection

Embarking upon the journey of pregnancy is akin to navigating a realm brimming with wonder, yet fraught with intricacies that often elude the unprepared. Among these, the influence of blood types is a tapestry woven with threads of crucial medical significance, as it underpins not only the health of the expectant mother but also the nascent life within her.

Our goal, then, is to traverse this landscape with care, ensuring that the delicate interplay of blood types between mother and child is understood and managed with the utmost of medical prudence. Through this guide, we will illuminate the path, equipping readers with the knowledge to foster a safe and healthy pregnancy, culminating in the joy of childbirth free from the shadows of preventable complications.

To embark on this journey, one must amass an arsenal of knowledge and resources. A thorough comprehension of one's own blood type and that of one's partner is paramount. Additionally, access to prenatal care with a healthcare provider versed in the complexities of blood type compatibility is indispensable.

Let us commence with a broad overview. The process is a tapestry of interconnected threads: understanding blood types and the Rh factor, recognising potential risks associated with incompatibility, and adopting proactive measures during pregnancy. Knowledge of prenatal testing and the administration of necessary interventions such as Rh immunoglobulin further fortifies this fabric.

With the stage set, we delve into the details, weaving through each phase with meticulous care. The human blood type is determined by the presence or absence of certain antigens on the surface of red blood cells—a genetic legacy bestowed upon us by our parents. In the context of pregnancy, the most pivotal of these antigens are the ABO blood group system and the Rh factor.

The ABO system categorises blood into four primary types: A, B, AB, and O, each with its unique antigenic profile. The Rh factor, on the other hand, denotes the presence (Rh positive) or absence (Rh negative) of the D antigen. When a discrepancy arises between the Rh blood types of a mother and her foetus—specifically, when a Rh-negative mother carries an

Rh-positive child—a vigilant eye must be cast toward the potential for Rh incompatibility.

What if the life within harbors an antigen foreign to the mother's bloodstream? Herein lies the risk of haemolytic disease of the newborn (HDN), a condition where the mother's immune system mounts a defensive onslaught against the foetal red blood cells. This can lead to serious complications, including anaemia, jaundice, or even heart failure in the foetus.

Let me offer you a pearl of wisdom, a lantern to guide you through this labyrinth. It is the vigilant monitoring of the mother's antibody levels and the timely administration of Rh immunoglobulin that can prevent the mother's immune response from turning against the foetus. This medication, given at around the 28th week of pregnancy and again after delivery, acts as a shield, safeguarding the unborn child from harm.

How does one ascertain that this protective bastion has been firmly established? Prenatal testing, including blood typing and antibody screening, becomes the compass by which we navigate these waters. Ultrasounds and amniocentesis provide a glimpse into the well-being of the foetus, ensuring that all is as it should be.

Yet, even the most carefully charted courses can encounter unexpected squalls. Should a Rh-negative mother inadvertently miss her Rh immunoglobulin treatment, or if an Rh incompatibility is diagnosed later in pregnancy, fear not. There are contingencies in place. Intrauterine transfusions and early delivery are among the harbor lights that guide us to safety.

In the vastness of this odyssey, remember that simplicity is your ally. Complex medical jargon is a tempest best left to those who sail the seas of medicine. Here, we seek to chart a course in clear, navigable waters, ensuring that each step taken is one of understanding and confidence.

In the symphony of life, the rhythm and cadence of a heartbeat are the most profound. Each chapter of this guide is composed with a mindful balance, mirroring the heartbeat's own dynamic rhythm—a testament to the life it seeks to protect.

At times, the voices of those who have journeyed before us echo in our minds, their experiences woven into the fabric of this guide. These quotations, these dialogues with the past, lend authenticity and depth to our narrative.

Show don't just tell, for the story of blood types and pregnancy is a tapestry rich in detail and nuance. Through anecdotes and vivid descriptions, we paint a picture that extends beyond the canvas of words, allowing the reader to see, to feel, to know the importance of this knowledge.

As we stand at the threshold of this new chapter, remember that blood ties, indeed, bind us together beyond the realm of race, often more than one might think. In the realm of pregnancy and childbirth, they are not mere characters in a tale of genetics but pivotal players upon who's understanding the health and well-being of both mother and child so significantly depend.

Rh Factor: A Deep Dive

Understanding the intricate tapestry of human biology requires a firm grasp of the threads that bind it together. Among these threads, the Rh factor emerges as a critical element, particularly in the context of pregnancy. A deep dive into the Rh factor reveals its potential to introduce complications, but also the measures that can be taken to mitigate them.

To navigate the complexities of the Rh factor, it is essential to arm oneself with a lexicon of key terms. Antigens, antibodies, Rh incompatibility, haemolytic disease of the newborn (HDN), Rh immunoglobulin, and prenatal testing are the signposts that will guide us through this exploration.

Antigens are substances that can elicit an immune response. They are like flags that adorn the surface of red blood cells, signalling the cell's identity. In the Rh system, the primary antigen of concern is the D antigen. If present, it renders the blood Rh positive; if absent, Rh negative.

Antibodies, in contrast, are the body's defenders. They patrol the bloodstream, seeking out foreign antigens to neutralise. When a Rh-negative individual is exposed to Rh-positive blood, their immune system may produce Rh antibodies, a process known as sensitisation.

Rh incompatibility occurs when a Rh-negative mother is pregnant with an Rh-positive foetus, inherited from an Rh-positive father. In such cases, the mother's body might produce antibodies against the foetus's Rh-positive cells, potentially leading to HDN.

Haemolytic disease of the newborn is a serious condition resulting from Rh incompatibility. It can cause the destruction of the foetus's red blood cells, leading to anaemia, jaundice, and in severe cases, heart failure or hydrops fetalis.

Rh immunoglobulin is a preventive medication administered to Rh-negative mothers. This medication contains Rh-positive antibodies that neutralise any foetal Rh-positive cells in the mother's bloodstream before her immune system can react to them.

Prenatal testing encompasses a suite of procedures to monitor the health of both mother and foetus throughout pregnancy. It includes blood typing, antibody screening, and in some cases, ultrasounds, and amniocentesis to assess foetal well-being and detect any signs of HDN.

Imagine a garden where plants of different kinds thrive together harmoniously. This is akin to a pregnancy where Rh factor incompatibility is managed successfully. In this garden, the vigilant gardener – the healthcare provider – uses the tools of knowledge and intervention to ensure that each plant, each life, grows without interference from preventable complications.

But what if, despite precautions, the immune system begins its silent march against the unseen guest? Picture a peaceful army suddenly recognising an invader, its soldiers rallying to defend their territory. This is the crux of Rh incompatibility, where the mother's body may perceive her own child's blood cells as foreign.

Does the infant, cocooned in the womb, sense the battle waged in its name? While the foetus remains unaware, the mother's medical team stands on guard. They utilise the Rh immunoglobulin as a peace envoy, offering a truce before the conflict escalates.

Occasionally, one may wonder, "What if the unseen dance of antigens and antibodies within me could affect my child?" This question pokes at the heart of why understanding the Rh factor is paramount for expecting mothers.

To truly embody the gravity of Rh incompatibility, consider the story of a tiny boat caught in a tempest. The mother's immune system is the storm, the Rh-positive foetal cells the boat. Without intervention, the boat risks being consumed by the waves.

The application of Rh immunoglobulin is akin to a lighthouse guiding the boat to safety, its beacon a symbol of protection against the storm. Through the administration of this crucial medication, a safe passage is assured for the tiny vessel.

In the realm of prenatal care, simplicity is the anchor that keeps us grounded. We steer clear of the tempest of medical jargon, opting instead for the calm waters of clarity and understanding. Through this approach, we ensure that each concept, each protective measure, resonates with the reader.

The rhythm of our narrative ebbs and flows, mirroring the cadence of a mother's heartbeat – steady, reassuring, a testament to the life it supports. Our prose dances with variety, employing the occasional one-line paragraph to underscore a pivotal point.

We draw upon the stories of those who have faced Rh incompatibility, their voices echoing through time. Their experiences, distilled into quotes and dialogues, enrich our understanding with authenticity and depth.

Rather than simply recounting facts, we strive to show the implications of the Rh factor through examples and anecdotes. Vivid imagery paints a picture that extends beyond the canvas of words, engaging the reader's senses and emotions.

In the grand narrative of pregnancy, the Rh factor plays a pivotal role, more critical than one might initially perceive. As we navigate the journey of understanding, we recognise that blood types, particularly the Rh factor, are not mere characters in a genetic story. They are key players whose influence on the health and well-being of both mother and child cannot be overstated.

Maternal-Foetal Harmony: Case Studies

In the quiet town of Edithvale, a coastal suburb where the scent of the ocean mingled with the bustling life of its inhabitants, a case unfolded that would underscore the paramount importance of blood type compatibility in maternal and foetal health. This was not merely an academic exercise for me, but a profound journey into the intricacies of human biology.

At the heart of this case were two main players: Emily, a first-time mother with the glow of expectant life upon her, and her unborn child, whose silent presence was the focus of our concern. Emily, a vibrant teacher with an infectious laugh, had O negative blood, the universal donor type known for its rarity and the absence of A, B, and Rh antigens.

The challenge emerged during a routine prenatal visit when blood tests revealed the formation of antibodies in Emily's blood—a stark indication that her foetus likely possessed a Rh-positive blood type, inherited from its father. This discovery set the stage for potential Rh incompatibility, a situation fraught with risk for the developing child.

To address this looming threat, the obstetric team deployed a meticulous strategy. Emily's blood work was monitored closely for rising antibody titters, and an ultrasound was ordered to check for any signs of foetal anaemia. Additionally, the medical team prepared for the possibility of an intrauterine transfusion, should the need arise to directly supply the foetus with compatible Rh-negative blood.

The results of this proactive approach were a testament to the power of medical intervention. Emily's pregnancy was closely managed, and through careful monitoring and timely treatment, the development of severe haemolytic disease in the newborn was successfully averted. The moments following the birth were marked by relief and joy, as the baby, though slightly jaundiced, was declared healthy and free from the feared complications of Rh incompatibility.

In the aftermath of this harrowing experience, reflection revealed broader insights. The case of Emily and her child exemplified the critical role that blood type compatibility plays in ensuring maternal and foetal well-being. It highlighted the importance of vigilant prenatal care and the necessity for expecting mothers to be aware of their Rh status.

Enhancing our understanding further, a visual aid, perhaps a chart or diagram, could have illustrated the mechanism of

Rh incompatibility and the medical interventions employed. Such aids would serve to demystify the process for future parents and medical professionals alike.

This singular narrative extended beyond the confines of one family's experience, connecting to the larger concept that blood types, often overshadowed by discussions of race or ethnicity, are indeed vital determinants of health in pregnancy. It is a reminder that the delicate balance of maternal-foetal harmony hinges not on visible traits but on the unseen molecular markers coursing through our veins.

As readers absorb the gravity of these revelations, one might ponder, "How often do we overlook the significance of our own blood type until it becomes a silent herald of potential strife?" This question invites contemplation on the everyday miracles of life that go unnoticed until challenged by the unexpected.

In our exploration of blood type compatibility and its implications, we have ventured through a landscape of medical knowledge, avoiding the thorny patches of complex jargon. Our language remained accessible, ensuring the essence of our message was not lost in translation.

The rhythm of our narrative was akin to the steady pulse of life, punctuated by moments of emphasis—a single sentence paragraph here, a direct question there—guiding the reader through the ebb and flow of the text.

We have not merely told the story; we have shown it through the lens of Emily's journey, her fears, her resilience, and the unwavering support of her medical team. We have painted with words the picture of a challenge faced and overcome, leaving an indelible mark on the canvas of our collective understanding.

In closing, the case studies presented herein serve as a clarion call to recognise the vital role blood type plays in the intricate dance of life. Let us continue to delve deeper into the wellspring of knowledge that shapes our understanding of maternal-foetal harmony, ever mindful of the delicate symphony that plays out within.

The Genetics of Motherhood: Beyond Blood

In the tender cradle of motherhood, where a woman's body becomes a nurturing vessel, the dance of genetics choreographs a complex interplay between her and her child. Here, in this sacred space, the conversation evolves, transcending the significance of blood types to encompass a broader genetic narrative. We embark on a journey through the labyrinthine pathways of DNA, where the genetic factors influencing the mother-child relationship are as diverse as the colours in a kaleidoscope.

Why juxtapose blood type with the vast expanse of genetic influences? The quest for understanding beckons us to explore this terrain, to unearth the insights buried within the confluence of genes and the roles they play in shaping the maternal bond. The purpose is clear: to illuminate the shadows, to extend our gaze beyond the crimson tide of blood type, and to embrace the rich tapestry of genetics that weaves together the story of mother and child.

As we delve into this exploration, we establish criteria for our comparison: the influence on prenatal development, the potential for health complications, and the implications for the emotional and physiological bonds between mother and offspring.

Blood type, with its direct implications for compatibility and transfusion reactions, shares common ground with genetic factors that also govern development and health outcomes. Both serve as silent sentinels, guiding the development of the foetus, with the potential to either harmonise with or challenge the maternal environment.

Yet, the contrast is striking. Where blood type compatibility is often distilled into a single concern—Rh factor or ABO discrepancies—genetic factors encompass a vast array of variables. From the nuances of inherited conditions to the subtleties of gene expression, each genetic marker carries its own story, its own potential impact on the maternal-child dyad.

Imagine, if you will, a visual aid: a double helix spiralling alongside a blood drop, each twist and turn of the helix, each molecule within the drop, representing the myriad factors at play. Such imagery serves to crystallise the complexity of this genetic symphony.

The analysis digs deeper. What do these comparisons reveal? They shine a light on the multifaceted nature of inheritance, on the layers of genetic interaction that can influence everything from a child's eye colour to their susceptibility to certain diseases. They underscore the reality that the genetic link between mother and child is not a single thread but a woven fabric, resilient yet subject to the patterns set forth by ancestral genes.

Is the relevance of these insights confined to the pages of textbooks or the walls of laboratories? Far from it. In a world where personal genomics and prenatal screenings are becoming increasingly commonplace, these revelations take on a new urgency. They compel us to consider not only how we prepare for potential health challenges but also how we understand the ties that bind us to our progeny.

One might ask: "In what ways do these genetic intricacies shape the daily realities of motherhood?" Indeed, the question invites us to reflect on the profound influence of genetics on the expectations, fears, and hopes that accompany the journey of bringing a new life into this world.

Our narrative wends its way through this genetic landscape with a commitment to clarity. Complex terms are distilled into simple language, ensuring that the profound significance of these genetic threads is not entangled in the web of jargon.

The rhythm of our prose mirrors the heartbeat of discovery—a staccato of short insights interspersed with the flowing melody of detailed exposition. Here, a one-line paragraph stands as a beacon, emphasising a crucial point; there, a dialogue between expectant mother and genetic counsellor breathes life into the abstract.

We do not merely inform; we evoke. Through anecdotes of genetic triumphs and the quiet strength of mothers facing genetic adversities, we paint a portrait of the indomitable spirit that defines the maternal experience.

Let us acknowledge the profound depth of the genetic well-spring that nourishes the roots of motherhood. As we continue to chart the course of this genetic odyssey, may we remain ever mindful of the intricate dance of life's code and the enduring bond it forges between mother and child.

Blood Ties That Bind: The Emotional Aspect

Embark on a journey where the most profound connections are illuminated not by the visible spectrum of our skins but by the unseen ties of our bloodlines. As you turn these pages, allow yourself to be drawn into an intimate world where the essence of kinship transcends the superficial and taps into the universal pulse of human emotion.

Within the heart of every mother beats a rhythm that resonates with her child, a symphony of shared blood that composes a song of life unlike any other. Here, in the sanctum of connected lives, you will discover the intricate dance of maternal love and genetic bonds—a dance that weaves the fabric of our beings into the tapestry of family.

Can you imagine the power of bloodlines that hold within them the secrets of our deepest affections? This tome reveals the hidden threads that bind us, the biological currents that drive the maternal instinct beyond the realm of conscious thought.

Perhaps you are sceptical, questioning the veracity of such claims. Rest assured, this scepticism will be met with a wealth of evidence, a series of revelations born from the confluence of medical science and the palpable experiences of motherhood. As a student of medical science, I weave together the empirical with the emotional, validating the profound connections that define our human condition.

Envision, for a moment, the transformation that awaits as you transcend the boundaries of conventional understanding. This is not merely a text; it is an invitation to witness the alchemy of genetics and emotion, to see the invisible fibbers that connect a mother to her child in a bond that is both biologically ordained and emotionally charged.

This book is a commitment, a pledge to guide you through the labyrinth of love and lineage. As you delve deeper, you will uncover the layers of maternal affection that are informed by the very substance coursing through our veins. This is where the power of knowledge meets the strength of feeling, where the science of blood types becomes a metaphor for the enduring connections of the human heart.

In the realm of motherhood, bloodlines are not mere identifiers; they are the harbingers of a love so potent that it transcends the boundaries of race and ethnicity. Whether you are a mother, a child, or simply a seeker of understanding, the revelations contained within these pages will redefine your perception of kinship.

With every chapter, you will be drawn into a world where stories of mothers and their children illuminate the emotional landscape of our shared humanity. You will encounter narratives that challenge the status quo, that offer a new lens through which to view the ties that truly bind us.

Are the emotional bonds of motherhood merely instinctual, or do they carry within them the weight of our genetic inheritance? This question will be explored through the vivid imagery of shared moments, the silent understanding that passes between a mother and her child with but a glance.

You are invited to engage with these pages, to question and to reflect. What does it mean to share a bloodline? How do the silent whispers of our genetic makeup inform the intensity of a mother's love? These are the queries that stir the soul and provoke a deeper appreciation for the connections that define us.

In crafting this narrative, adverbs and adjectives are used with precision, each chosen to enhance rather than overshadow the core message. The power of this story lies not in its ornate language but in its ability to capture the essence of maternal bonds with clarity and truth.

Occasionally, a one-line paragraph will punctuate the flow, a solitary sentinel standing guard over a truth too significant to be lost in a sea of words. These moments are crafted to resonate, to echo in the chambers of your heart long after the page is turned.

Above all, this work is accessible, written in a language that embraces the reader, regardless of their scientific acumen. It is a bridge between the expert and the layperson, a conduit for shared understanding in the realm of blood and feeling.

The rhythm of the prose ebbs and flows, mirroring the cadence of life itself. In these patterns, you will find a reflection of the heartbeat of motherhood, a pulse that quickens with joy and steadies in the comfort of belonging.

Through quotations and dialogues, you will hear the voices of those who live this truth, who breathe life into the science with their experiences and emotions. In these exchanges, the abstract becomes tangible, the theoretical personal.

This book does not merely tell; it shows. It invites you to walk alongside mothers as they navigate the waters of genetic destiny, as they confront the challenges and revel in the triumphs that bloodlines bring.

Prepare to be transformed, to emerge from this reading with a new understanding of what it means to be connected by blood. This is a journey of discovery, a path that leads to the heart of what it truly means to belong. Welcome to "Blood Ties That Bind: The Emotional Aspect," where the depth of maternal love and the power of bloodlines converge to reveal the most fundamental truths of our existence. Let us begin.

CHAPTER

10

Racial Constructs: A Historical Dissection

Human history is a tapestry, woven with the variegated threads of culture, belief, and biology. The notion of race, a thread so deeply embedded in the fabric of society, bears a tale that is both intricate and fraught with contention. But whence did this idea originate, and how has it transformed through the ages to become the complex and divisive concept it is today?

In the earliest records of human civilisation, there is scant mention of race as we understand it. The inception of the idea can be traced to a time when societies began to encounter others with marked differences in appearance, language, and customs. It was in these nascent interactions that the seeds of racial thought were sown.

Ancient civilisations, such as the Greeks and Romans, classified people based on their geographical origins and traits, but these categorizations were fluid and not rigidly bound to notions of superiority or inferiority. A Greek might speak of the "barbarians" beyond their lands, yet this was more a distinction of culture and language than of immutable biological difference.

The tapestry darkens as we traverse the timeline. With the Age of Exploration, European explorers encountered indigenous peoples across the globe. Here, we find a pivotal milestone: the pseudo-scientific classification of humans into distinct races, a practice that became entrenched with Carl Linnaeus's 18th-century taxonomy, which sought to categorise the natural world.

Imagine, if you will, an engraving from that era—depicting the "five races" of mankind. Such images, though now recognised as deeply flawed, were once considered emblematic of scientific thought. They provided a visual reinforcement to the hierarchy that was beginning to take shape in the collective mindset, influencing perceptions across continents.

The concept of race evolved differently across cultures. While Europeans developed a hierarchy that placed themselves at the pinnacle, other cultures maintained more complex, often less rigidly hierarchical views. In many regions of Africa and Asia, for example, in Ethiopia, ethnicity and tribal affiliations played a more significant role than skin colour in social stratification.

What, then, of the modern interpretations of race? Twentieth-century genetic research has revealed that the biological basis for race is virtually non-existent—humankind is far more similar than different at the genetic level. Yet the social and political realities of race remain potent forces, shaping identities and experiences in profound ways.

Have we not witnessed, in recent years, a paradigm shifts in racial discourse? Movements for social justice have brought to light the enduring impact of racial constructs, challenging us to confront uncomfortable truths and work towards a more equitable society.

It is impossible to discuss the evolution of the race without addressing the controversies that have erupted over time. The scientific racism of the 19th and early 20th centuries, for example, provided justification for colonialism, slavery, and eugenics. These dark chapters in human history force us to reckon with the consequences of our beliefs and the knowledge we choose to propagate.

The narrative of race is one replete with turning points—moments that have demanded a re-evaluation of long-held assumptions. Consider the Civil Rights Movement in the United States, a struggle that not only changed laws but also challenged the very foundation of racial thinking.

How can one encapsulate the essence of such a complex history? Perhaps the answer lies not in the pursuit of absolutes but in the recognition of the fluidity and interconnectedness of human narratives.

As we chart the course of this history, we must ask ourselves: What have we learned from the past, and how will we allow that knowledge to shape our future? Will we cling to antiquated notions, or will we strive to build a world where the content of one's character truly matters more than the colour of one's skin?

The journey through the history of race is a labyrinthine passage, one that continues to evolve as we expand our understanding and challenge our prejudices. This is but a glimpse into that journey, a primer for the profound exploration that lies ahead. And as we turn the page, we carry with us the hope that through understanding our past, we may forge a future defined not by division, but by unity in our common humanity.

Blood Types and Racial Categories: A Scientific Mismatch

In the quest to unravel the complex narrative of human categorisation, we find ourselves at the crossroads of biology and society. On one path lies the scientifically measurable, the tangible essence of our being—the blood that courses through our veins, identifiable by type. On the other, the social construct of race, an idea that has shaped civilisations and seeded division. What insights can be gleaned from examining these two distinctive markers? Why does one, rooted in clear genetic expression, hold less sway in our social interactions than the other, which science has struggled to define?

Blood types, a biological reality, are determined by the presence or absence of certain antigens on the surface of red blood cells—a simple, yet critical factor that can mean the difference between life and death in a medical emergency. These types, known universally as A, B, AB, and O, intermingle and manifest across the human population without regard to borders, language, or culture. They represent a universal commonality, an aspect of human life that unites us in our vulnerability and our need for one another—after all, a transfusion knows no prejudice.

Race, by stark contrast, has no such clear-cut genetic foundation. While humans are adept at discerning visual and cultural differences, genetic studies have shown that the variation within so-called racial groups is often greater than that between them. The concept of race, it seems, is a social invention, a way of categorising people that has shifted and changed throughout history, often with devastating consequences.

The criteria for comparison between blood types and racial categories are fundamentally different. Blood types are classified based on clear, observable, and testable criteria. Race, on the other hand, is often based on subjective perceptions and historical contexts that can vary widely between societies and over time.

Yet, is there not a strange duality here? Blood types can be seen as a metaphor for human diversity—each type unique, yet each person's blood performing the same vital functions. Similarly, the human tapestry is rich with cultural and physical diversity, but at our core, we are united in our humanity.

The differences are stark when viewed side by side. Blood type, regardless of category, carries no inherent value judgment—it does not influence one's potential, intelligence, or worth. However, the racial category one is perceived to belong to can have profound effects on their life experiences, opportunities, and how they are treated by society.

Consider the visual aids of blood type charts in a classroom, displaying compatibility for donations and transfusions. There is no hierarchy—just a practical guide to saving lives. Now envision the historical illustrations of racial hierarchies, steeped in bias and pseudoscience, which have been used to justify discrimination and inequality. The contrast could not be more pronounced.

What do these comparisons reveal? They lay bare the fact that while blood types are an essential, objective aspect of our biology, race is a subjective social construct that has been used to otherise and oppress. The implications of this are broad, touching on issues of identity, health care, and social justice.

In the contemporary context, the relevance of blood types has taken on new dimensions with the advent of personalised medicine, where treatments can be tailored to one's genetic profile. Race, meanwhile, remains a contentious and powerful force, influencing everything from politics to personal identity.

Does this not prompt a reflection on our values and priorities? Why do we cling to the concept of race, with its murky definitions and painful legacy, when more precise and scientifically sound methods of understanding human diversity exist? Can we envision a future where the content of one's character and the compatibility of their blood hold greater significance than the colour of their skin or the ancestry they claim?

In the end, the discourse on blood types and racial categories challenges us to consider the foundations of our beliefs and the structures of our societies. It beckons us to ask: Could the recognition of our shared biological vulnerabilities and mutual dependencies foster a more empathetic and united world?

The narrative of race, a complex and enduring thread in the human story, is slowly being unravelled by the steady hands of science and the determined hearts of those who seek equity and truth. As we continue to write this book, we carry with us the hope that by understanding our similarities and respecting our differences, we might weave a new tapestry—one that celebrates the multifaceted beauty of all humankind.

Colonialism and Blood: The Untold Story

In the crucible of history, where the past is prologue and the present is but a fleeting moment, we cast our gaze back to the days when empires sprawled like great beasts across the continents. Their tendrils, wrapped in the guise of trade and exploration, reached far and wide, driven by a hunger for resources, land, and dominion. Within this age of conquest and dominion, a less told narrative simmered beneath the surface, one that hinged not on the colour of one's skin but the essence of their blood.

In the twilight of the 19th century, as the sun never set on the British Empire and the Scramble for Africa carved up a continent, a discovery in a Viennese laboratory would quietly take its place among the milestones of medical science. Karl

Landsteiner, a name now etched in the annals of history, unveiled the ABO blood group system, a key that would unlock the mysteries of human transfusion and compatibility.

Yet, what did this mean in a world where the lines drawn by colonial powers were of a different nature? They were etched in soil and sand, drawn through tribal lands and ancient kingdoms, heedless of the cultural and societal ruptures they would cause. How did Landsteiner's pioneering work on blood types play into the grand chessboard of colonial ambition and racial ideology?

Was it mere coincidence that the very fluids which could save lives in the surgical wards of Europe's grand cities were coursing through the veins of those deemed 'lesser' by the arbiters of empire? Could the science of blood types have offered a universalising thread, binding humanity in its shared biology, in stark contrast to the divisive stratagems of imperial rule?

In time, the tumult of world wars would demonstrate the critical importance of blood type understanding. On battlefields soaked with the blood of countless nationalities, the soldier's life might hinge not on the allegiance on their uniform but the type of blood that could be transfused into their veins.

As our journey brings us from the past to the present, we must ask: What relevance does this knowledge hold today? The tendrils of colonialism may have withered, or mutated into neo-colonial forms, but the lessons of our shared biology remain as pertinent as ever. The blood banks of the modern world, stripped of prejudice, do not discriminate based on the donor's race or ethnicity. Blood type, not skin colour, dictates the flow of life-giving fluids from one person to another.

Why then, does the spectre of race still loom large over our contemporary landscape? Has the scientific community not provided us with a clearer, more equitable lens through which to view our shared humanity?

Consider the implications for a moment. In a world where blood type understanding mattered beyond race, what changes might unfold? Would the disparities in healthcare, the biases in blood donation practices across different communities, be addressed with renewed vigour?

And yet, the question stands before us, stark and unyielding: Are we prepared to embrace this aspect of our biology as a foundational principle of equality, or will the old constructs, born of a colonial past, persist in their hold over the collective psyche?

As we segue into the story of blood in the context of colonialism, we delve deeper into the complex tapestry of human interaction. Here, we will explore tales of exploitation and enlightenment, of scientific advances wielded as tools of oppression and as instruments of liberation.

The untold story of colonialism and blood awaits, a narrative steeped in the ironies of history, where the very substance that unites us, all was once overshadowed by the ideologies that sought to divide us. Let us now begin to unravel these threads, to reveal the rich, red blood that might yet write a new chapter in the story of humankind.

Race in Science: From Pseudoscience to Enlightenment

As the dawn of the 20th century approached, the world found itself on the precipice of monumental change. The understanding of human blood—its types, its properties, and its profound implications—was poised to alter the course of medical science and societal perceptions.

The initial breakthroughs had been made in the waning years of the 19th century, but the true impact of these discoveries would only become apparent as the new century unfolded. With the ABO blood group system identified, the foundation was set for a deeper appreciation of human physiology that transcended superficial differences.

The earliest origins of this enlightenment lay in the hands of pioneers like Karl Landsteiner, but his was not the only voice in the choir of scientific progress. Others would follow, building upon his work, each contributing a verse to the evolving song of understanding. Yet, amidst this chorus of advancement, the world grappled with the shackles of racial pseudoscience, a doctrine that had long sought to categorise and divide humanity based on external features and unfounded hierarchies.

Throughout the early 20th century, major milestones marked the path from ignorance to insight. The discovery of the Rh factor by Landsteiner and Alexander Wiener in 1937 expanded the horizons of blood type research, revealing yet another layer of complexity within human blood. This revelation held profound implications, not only for the burgeoning field of transfusion medicine but also for the slowly turning tides of social consciousness.

Do you ever pause to wonder how such a discovery might have shifted the societal landscape if its full potential had been recognised sooner?

Visual aids from the time, like charts of blood type distributions and diagrams of the Rh factor inheritance, serve as silent testaments to the progress of science. They remind us that beneath the skin, the blood coursing through our veins is a testament to our shared humanity.

As this knowledge spread, variations in its reception and application emerged across cultures and regions. In some parts of the world, blood type became a curious cultural fixation, influencing everything from personality assessments to matchmaking. In others, it remained a purely medical consideration, vital for ensuring the safety of transfusions and the health of mothers and their newborns.

The modern interpretations of these discoveries have been varied and complex. In recent years, the significance of blood types has permeated popular culture, with books and articles purporting to link blood type to diet, temperament, and more. While these claims often lack rigorous scientific backing, they point to an ongoing fascination with the idea that our biology can reveal deeper truths about who we are.

Yet, as we have journeyed through this evolution of understanding, we have also encountered challenges and controversies. The dark spectre of eugenics, which sought to apply genetic principles to engineer a 'better' human race, is a grim chapter in our history. It is a stark reminder that scientific knowledge, devoid of ethical considerations, can lead to unspeakable atrocity.

Would it surprise you to know that even in the enlightened era of the 21st century, the echoes of these misdeeds still reverberate in the halls of modern science?

And what of the turning points we have witnessed? The universal donor, type O-negative blood, once a mere footnote in the annals of science, has become a symbol of life-saving altruism. The realisation that anyone, regardless of race or ethnicity, could provide this precious resource has quietly underscored the folly of racial divisions.

One cannot help but ponder: what if the commonality of our blood had been the guiding principle of our societies from the outset? How different might our world look today?

In conclusion, the journey from racial pseudoscience to a more enlightened understanding based on blood type is one of both triumph and tragedy. It is a tale that continues to unfold, reminding us that the path to enlightenment is not a straight line but a road fraught with turns and obstacles.

We stand at a crossroads, with the power to use our knowledge for the betterment of all. May we choose wisely, ever mindful of the blood that unites us, ever vigilant against the remnants of ignorance that once sought to divide us.

Let us now turn the page, leaving behind the outmoded notions of the past, and step forward into a future where blood ties extend beyond the realm of race, and humanity is celebrated in all its diverse unity.

Racial Constructs in the Modern Age: Where Do We Stand?

In the tapestry of humanity, a myriad of threads weaves together to form the picture of our shared existence. Each thread represents a different aspect of who we are, from the languages we speak to the cultures we celebrate. But amongst these, one thread has historically been pulled taut, often to the point of fraying the fabric itself: the concept of race.

Race, as it has been understood, is a social construct with no basis in scientific fact. It is a classification system that was created to categorise people based on physical characteristics, such as skin colour and facial features. While these attributes are genetically inherited, they represent only superficial variations among populations and do not reflect distinct biological differences.

The concept of race has been used to justify a range of inequalities and injustices throughout history. Its roots can be traced back to times when imperial powers colonised foreign lands, imposing a hierarchy that placed certain groups above others. These racial classifications have had profound and enduring impacts, influencing socioeconomic status, access to education, and healthcare outcomes among many other facets of life.

Yet, delve deeper into the realm of genetics and a different picture emerges—one where the aforementioned physical traits are revealed to be mere surface-level differences. The Human Genome Project, completed in the early 21st century, confirmed that all humans share 99.9% of their DNA.

What's more, this genetic similarity transcends all so-called racial boundaries.

Within this broader framework, blood types present a compelling counter-narrative to racial distinctions. Unlike race, blood types are defined by specific markers on red blood cells and are critical for medical purposes, such as blood transfusions and organ transplants. These markers must be compatible between donor and recipient for such procedures to be successful, and this necessity disregards any notions of racial divide.

Consider the real-world implications of blood type compatibility. A patient in dire need of a transfusion cannot afford the luxury of discrimination; the only criterion that matters is the presence of a compatible blood type. The search for a matching donor thus becomes a unifying quest that transcends racial and ethnic lines, illustrating our biological interconnectedness.

However, common misconceptions continue to muddy the waters. Some people still cling to outdated beliefs about the superiority or purity of certain races, despite the scientific evidence to the contrary. Misinterpretations of genetic research can fuel these erroneous ideas, leading to a resurgence of racial pseudoscience in various guises.

Are we not, then, obligated to confront these falsehoods with the truth that our most fundamental life force—our blood—is the great equaliser? Could this recognition not be the catalyst for a shift in perspective, one that sees us prioritise our shared humanity over divisive constructs?

What if, instead of allowing the abstract and arbitrary lines of race to dictate our societal structures, we embraced the concrete and critical significance of blood types? Imagine the potential for a society where individuals are valued not for the colour of their skin but for the life-saving compatibility of their blood. This vision is not merely utopian; it is grounded in the very essence of our biology.

The conversation around race and blood types is, admittedly, complex. Blood types themselves are not immune to patterns of distribution across populations, which can sometimes overlap with ethnic groups due to shared ancestry. It is crucial to approach these discussions with nuance, acknowledging the complexities without reinforcing the erroneous concept of race as a biological reality.

In the face of these intricacies, we find ourselves at a juncture in history. Will we continue to be ensnared by the archaic and artificial divides of race, or will we rise to embrace a more enlightened perspective? Can we uphold the truth that, when it comes to the most essential aspects of life and health, blood ties beyond race?

As we navigate the modern age, let us hold fast to the understanding that our shared blood is a testament to our common humanity. Let us wield this knowledge as a tool to dismantle the remnants of racial constructs and build, in their place, a society founded on the principles of equality, unity, and mutual respect.

For in every drop of blood lies the potential to heal or to harm. The choice of which path we follow rests with us. Let us choose the path of healing, guided by the wisdom of science and the compassion of our shared human spirit.

CHAPTER

11

The Global Blood Type Phenomenon

In the quiet laboratories of early 20th-century Vienna, a discovery unfolded that would ripple through time, shaping cultural conversations and social dynamics across continents. Karl Landsteiner, a scientist whose curiosity matched the fervour of his discipline, revealed to the world the existence of blood types. His monumental work laid the foundation for understanding human blood groups, an insight that would transcend mere medical curiosity and seep into the fabric of popular culture.

The journey Landsteiner embarked upon at the dawn of the 1900s would unravel the mystery of why some blood transfusions were successful while others ended in tragedy. His revelation of the ABO blood group system brought clarity to this life-and-death puzzle. This momentous discovery not only revolutionised medical transfusions but also unwittingly set the stage for blood types to become a lens through which society would view personality, compatibility, and identity.

As the decades passed, the world witnessed upheavals and transformations that would challenge the status quo and redefine cultural norms. The aftermath of two World Wars, the rise of global communication, and the advent of mass media turned society's gaze to new forms of understanding human nature. It was within this crucible of change that blood types emerged as an unexpected social compass.

Fast forward to today, and one can hardly escape the intrigue surrounding blood types. Have you ever wondered why certain forms of entertainment, particularly in East Asia, place such emphasis on a character's blood type? Why do dating profiles in some countries tout blood type compatibility as a fundamental criterion for potential partners? What does it mean when a celebrity's temperament is casually attributed to being type A, B, AB, or O?

The importance of historical understanding is paramount when seeking answers to these contemporary curiosities. Without the context of where and how the fascination with blood types began, one is left navigating a sea of assumptions without a compass. The significance of Landsteiner's work is not merely in its scientific merit but also in its unexpected cultural legacy, one that continues to influence how individuals perceive themselves and others.

So, how did this medical concept evolve into a pop culture phenomenon? The answer lies in the universal human quest for identity and the need for a sense of belonging. Blood types, with their seemingly enigmatic implications for personality and health, offer a unique form of self-discovery and community. In Japan, for instance, the concept of ketsueki-gata has become so ingrained that people may turn to blood types to navigate friendships, work relationships, and love life. Books, television shows, and even video games often feature characters whose blood types are indicative of their personalities, as if one's essence could be distilled into a single letter of the alphabet.

Consider, then, the power of a simple biological trait to captivate the imagination of millions. A whisper of one's blood type carries with it a symphony of presumed traits: type A's meticulous nature, type B's free spirit, type AB's complexity, and type O's leadership. But is there truth to these stereotypes, or are they but a collective myth we choose to entertain?

Why does this matter now, at a time when the world grapples with pressing global challenges? The answer is subtle yet profound. In understanding the cultural significance of blood types, we unlock a deeper comprehension of human behaviour and social dynamics. We learn about the stories we tell ourselves to create order in a complex world. We gain insight into how seemingly scientific concepts can take on lives of their own, shaping how we interact with one another.

As we delve deeper into the influence of blood types on pop culture, let us not forget the rich history from which this fascination springs. It is a narrative marked by scientific brilliance, marred by periods of misuse and misunderstanding, and ultimately, embraced by millions as a source of entertainment, identity, and connection. This book is an invitation to explore that narrative, to understand the power of a discovery made over a century ago, and to recognise its enduring impact on how we see ourselves and each other in the vast tapestry of humanity.

In the chapters that follow, we will traverse the globe, examining the unique ways blood types have permeated the zeitgeist of various cultures. We will meet individuals for whom these categories serve as social currency and dissect the industries that profit from this global fascination. Through this journey, we will seek to answer the pressing question: How have four simple letters come to hold such sway over the human heart and mind?

Blood Type-Based Products and Services

In the bustling metropolis of Tokyo, where neon signs cast a perpetual glow and the rhythm of life never skips a beat, there exists a curious market, one that caters to a biological signature etched in the very cells of its patrons—their blood types. This market is not confined to the vibrant streets of Japan's capital; it stretches across oceans, manifesting in various forms and capturing the imaginations of consumers worldwide. Here, in this intricate dance of commerce and culture, we find our setting.

At the heart of this phenomenon are the enterprising individuals and corporations that have recognised a lucrative opportunity. They are the main players in our story, the architects of a niche industry that has weaved the concept of blood type into the fabric of everyday life. From personalised dietary supplements to beauty products tailored to the minutiae of one's blood group, these entities have transformed a scientific classification into a cornerstone of individual identity.

The challenge they face is multifaceted: How to create products and services that not only resonate with the public's fascination with blood types but also stand up to scientific scrutiny and ethical considerations. The allure of customisation is potent, and the promise of a product "just for you" based on your blood type is enticing. Yet, beneath this veneer of personalisation lies a complex web of questions about the validity and implications of such offerings.

The approach to addressing this challenge is as varied as the products themselves. Strategies range from leveraging traditional beliefs about blood type personality traits to employing cutting-edge biotechnology to craft items purportedly optimised for one's physiological needs. Marketing campaigns are carefully crafted, blending science, superstition, and a dash of mystique to captivate their audience.

The results of these endeavours are mixed, with some products gaining cult-like followings while others fade into obscurity. Data on sales and customer satisfaction are guarded like state secrets, yet anecdotal evidence suggests a persistent demand for these customised goods. Blood type diets, for instance, continue to find adherents despite a lack of conclusive scientific evidence supporting their efficacy. Meanwhile, blood type matchmaking services boast success stories of couples brought together by the supposedly harmonious compatibility of their haemoglobin.

Upon reflection, this case study offers a rich tapestry of insights. It reveals a world where the desire for identity and belonging intersects with the power of modern marketing. It calls into question our willingness to embrace pseudoscience when it promises a sense of individuality and control. It also raises ethical questions about the commodification of our biological makeup and the potential reinforcement of stereotypes.

As we connect these specifics back to the overarching theme of our book, we find that the commercialisation of blood types is but one example of how we, as a society, seek to categorise and understand ourselves. It is a microcosm of the broader human narrative, where science and culture intertwine in unexpected and often inexplicable ways.

Let us pause and ponder: If the market for blood type-based products and services is a reflection of our collective consciousness, what does it say about our quest for identity in the modern world? Are we searching for meaning in the right places, or are we simply buying into a narrative that has been cleverly sold to us?

In the chapters that follow, we will delve deeper into the heart of this question, examining the implications of these practices on individual and societal levels. We will explore the intersection of commerce, science, and identity, and consider whether our preoccupation with blood types is a harmless fascination or a symptom of a deeper longing in the human spirit.

It is my hope, as a medical science student, that the journey through these pages will not only inform but also provoke critical thought and dialogue. For in the end, it is through questioning and seeking understanding that we truly learn about ourselves and the world we inhabit.

Blood Type Discrimination: A Hidden Bias

In the intricate fabric of society, where myriad threads of diversity intertwine to form the tapestry of humanity, there lurks a lesser known but equally insidious form of discrimination. This is not the oft-discussed discrimination of race, gender, or creed, but rather, one that flows within the very lifeblood of individuals—the discrimination based on blood types.

Blood types, though invisible to the naked eye, carry with them a baggage of biases that many are unaware of. This book ventures into the heart of this issue, unearthing the instances where an individual's A, B, AB, or O blood group has unjustly dictated aspects of their lives, opportunities, and societal roles. At its core, aims to illuminate the shadows where this form of discrimination resides, compelling readers to acknowledge and address the silent prejudices that permeate our social fabric.

In a world that prides itself on progressive thought and inclusivity, the notion of blood type discrimination may seem archaic, if not entirely fictitious. Yet, the impact of such biases is starkly real, affecting countless individuals in both overt and subtle ways. It is a phenomenon that has seen people judged not for the content of their character, but for the antigens on their red blood cells.

Why should it be that a person with type O blood is preferred for certain roles, deemed more reliable or dynamic, while those with type A blood are stereotyped as meticulous to a fault, or type B as indifferent and selfish? These stereotypes, grounded more in folklore than in fact, have led to life-changing decisions: from employment opportunities to romantic entanglements, even to the unsettling realms of eugenics and forced sterilisation in darker chapters of history.

To personalise the issue, consider the story of Hina, a vibrant young professional in Seoul, South Korea. Despite her qualifications, she found herself repeatedly passed over for promotions. The reason, whispered amongst colleagues as if it bore the weight of scientific truth, was her B blood type—supposedly indicative of a lackadaisical and unpredictable nature. Hina's tale is not an isolated incident; it is a narrative replicated in various forms across cultures and continents, a testament to the pervasive reach of this hidden bias.

The stakes in this silent ongoing prejudice are high. When we allow such unfounded judgments to influence our perceptions and decisions, we not only do a disservice to individuals like Hina, but we also stifle the potential within our communities. The diversity of blood types, much like the diversity of backgrounds, can be a source of strength and innovation rather than division—if only we learn to look beyond these archaic labels.

As we journey through the pages of this book, we seek not merely to present a litany of grievances but to chart a course towards understanding and change. We will explore the origins of these beliefs, dissect the pseudoscientific claims that bolster them, and highlight the voices of those who advocate for a shift in perception. Through interviews, case studies, and a critical examination of cultural narratives, we will unravel the complex interplay between blood type and identity.

Could it be that our enthrallment with the notion of blood types as determinants of personality and compatibility is but a mirror, reflecting our innate desire to categorise and simplify the complexities of human nature? Or perhaps, it is a manifestation of our yearning for a sense of belonging within the myriad communities that make up our social existence.

In confronting these questions, serves as a clarion call to readers. It is an invitation to challenge the status quo, to question the underlying assumptions that govern our interactions, and to take an active role in fostering a society that judges individuals by their actions and character, rather than the superficial markers found in their blood.

As we forge ahead, let us bear in mind the power of awareness and education. For it is only through shedding light on the dark corners of discrimination that we can hope to eradicate it. In this endeavour, seeks to be not just a book, but a beacon—a guide towards a more equitable and enlightened world where blood tie extend beyond the colour of one's eyes.

The Blood Type Community: Unity and Division

In the pursuit of unravelling the peculiarities of human categorisation, one stumbles upon a concept as intriguing as it is divisive—the community formed around blood types. This concept, while not initially self-evident, becomes a lens through which we view a unique form of social stratification.

To elucidate this phenomenon, we must first define what is meant by a 'blood type community.' Simply put, it is a social grouping, often based on the shared characteristic of having the same blood type, where members believe they possess common traits or are subject to similar life experiences due to this biological factor. Across different cultures, these communities have taken on identities of their own, attributing to each blood type a set of personality descriptors.

Delving into the key elements of these communities reveals a complex tapestry of beliefs and interactions. For instance, in Japan, the notion of 'ketsueki-gata' has woven itself into the fabric of social life, where blood types are thought to predict personality and compatibility in relationships, employment, and friendships. It's not uncommon for dating profiles to list blood types alongside hobbies and interests, or for job postings to suggest preferred blood types for applicants.

While this concept may seem modern, its roots can be traced back to early 20th-century eugenics movements, where blood type was one of the many factors used to promote ideas of superiority and inferiority among different groups of people. Thankfully, the world has largely moved past such abhorrent practices, but the remnants of such thinking persist subtly in the blood type communities of today.

When contextualised within a broader framework, these communities can be seen as microcosms of human desire for belonging and understanding. Just as people have historically grouped themselves by clan, nationality, or religion, blood types have become yet another axis along which communities form and assert their identities.

Real-world applications of the blood type community concept abound. A person with type O blood might be sought after in emergency situations for their universal donor status or might feel a sense of camaraderie with others of the same type. Conversely, those with type AB blood, the universal recipients, may find a unique bond in their rarity, as this is the least common blood type globally.

However, common misconceptions about blood type communities can mislead and perpetuate stereotypes. It's crucial to understand that there is no scientific evidence linking blood types to personality traits. The belief that type A individuals are inherently organised, or that those with type B blood are inherently selfish, is unfounded and can lead to unjust judgments and discrimination.

What, then, can be said of the real impact of these communities on the individual and society? Do they unite or divide us? The answer lies not in the blood running through our veins, but in the way we choose to interpret and act upon these classifications.

One's blood type, shrouded in mystery and misconceptions, says nothing of one's character, dreams, or capabilities. One might ponder, "What if we shed these labels, just as we strive to shed those of race and gender?" Would our social landscapes become more harmonious, or would we simply seek out new methods of division?

Let us consider the story of Emeka, a man with AB negative blood, who often felt like an outsider in his community due to the rarity of his blood type. Emeka's experience highlights the potential for isolation within these blood type communities, as well as the unity that can emerge when individuals with shared experiences connect.

Emeka's narrative is not just his own; it resonates with a universal theme of seeking connection in a world preoccupied with categorisation. The blood type community, therefore, serves as a double-edged sword—capable of forging bonds over shared biological traits, yet equally capable of slicing through the fabric of unity with the sharp edge of exclusion.

By integrating these elements, we can construct a nuanced view of the impact of blood type communities. They can be sources of support for some, while for others, they represent another boundary to be overcome in the quest for a more inclusive society.

In conclusion, is not just a study of groups formed on the basis of biology; it is a reflection on the human condition. It is a call to seek connections based on deeper, more meaningful attributes, and to recognise the diversity within us as a collective strength. May we all strive for a world where the content of one's character eclipses the label of one's blood type, and where unity is not dictated by antigens but by our shared humanity.

Blood Types and Global Health: A Public Perspective

As we delve into the intricacies of global health, we uncover a narrative that transcends the conventional discourse on race and ethnicity. It's a narrative that places a spotlight on a less explored, yet equally significant aspect of our biology: blood types. This vital fluid coursing through our veins is more than a lifeline—it is a key player in the realm of public health and global health initiatives.

In the grand theatre of global health, blood types take centre stage, not just as a medical curiosity, but as a pivotal factor in life-saving procedures and disease susceptibility research. The assertion here is bold yet underappreciated: blood types do, indeed, matters beyond the realm of race when it comes to certain global health policies and medical emergencies.

The grounding pillar of this claim rests on the universal need for blood transfusions. During a crisis, the difference between life and death can hinge on the availability of compatible blood. Take, for instance, the O negative blood type, the universal donor whose blood can be transfused into any patient in dire need. This type's paramount importance is evidenced by the prioritisation of O negative individuals in blood donation campaigns worldwide.

Digging deeper, we uncover the compelling narrative of how blood type distributions can influence public health strategies. In malaria-plagued regions, for example, there's a fascinating correlation between the prevalence of the O blood type and a reduced risk of severe malaria, suggesting a natural selection advantage. Such concrete evidence underscores the critical role that understanding blood type distributions can play in tailoring health initiatives to specific populations.

However, the narrative is not without its twists. Some may argue that focusing too heavily on blood types could overshadow other vital aspects of healthcare, such as socio-economic determinants and access to medical resources. While this viewpoint holds merit, it does not negate the importance of blood type considerations; rather, it calls for a more holistic approach to public health policies.

In response to such counterarguments, it is essential to clarify that advocating for blood type-focused health strategies does not imply a single-track mindset. Instead, it is about integrating this knowledge into a broader, multifaceted approach to health. Blood type information can enhance, not detract from, the effectiveness of global health initiatives when used in concert with other demographic and environmental data.

Furthermore, recent research reveals intriguing links between blood types and certain diseases, adding another layer of complexity to the narrative. For instance, studies have hinted at variations in COVID-19 infection rates and severity among different blood types, with types A and AB possibly facing a higher risk of severe outcomes. While these findings are preliminary and require further investigation, they underscore the potential of blood type data to inform public health responses in the face of pandemics.

To crystallise our understanding, consider the global response to the COVID-19 pandemic. Imagine if, armed with robust data on the relationship between blood types and the virus, health officials could prioritise vaccinations or tailor public health messages to those most at risk. The potential for such targeted strategies is immense, demonstrating the pragmatic value of blood type considerations in global health planning.

The assertion that blood ties extend beyond the realm of race in the context of global health is not only substantiated by concrete evidence but also carries profound implications for public health policies. This perspective beckons us to rethink our health strategies, to acknowledge the silent yet powerful influence of our blood types, and to champion a more inclusive and effective approach to global health. Let this be a clarion call—a call to recognise the significance of this biological factor that unites us in our diversity and holds untapped potential for enhancing the well-being of all.

CHAPTER 12

The Future of Blood: Transcending Race

Embark on an odyssey of discovery that will shatter your preconceptions and unveil the intricate tapestry of life that blood types of weaves. This book is not just a trove of information; it's a key to unlocking a new perspective on humanity.

Delve into the pioneering science that is poised to challenge the very foundations of medicine as we know it. As a dedicated medical science student, I have meticulously compiled ground-breaking research and innovative methodologies that will illuminate the path forward in the field of blood type research. Herein lies a confluence of genetics, anthropology, and immunology, seamlessly intertwined to unravel the enigmatic influence of blood types on our lives.

Perhaps you are sceptical. Why should blood types warrant such attention? How could they possibly be more significant than the visible traits that have divided humans for millennia? Consider this: every drop of blood coursing through our veins is a microcosm of history, a bearer of secrets waiting to be decoded. As you journey through these pages, your doubts will dissipate like mist in the morning sun, replaced by a sense of awe at the profundity of our biological heritage.

Imagine a world where the nuances of your blood could predict health outcomes, foretell responses to diseases, and tailor medical treatments with unprecedented precision. This is not mere fantasy; it is the imminent reality that this book will guide you towards. You will witness the transformative power of understanding blood types, not just as a static classification, but as a dynamic and critical factor in personalised medicine.

By turning the final page, you will not leave the same as you arrived. Your perception of race, heritage, and health will have undergone a metamorphosis, driven by the life-changing revelations contained within this exploration. The value of this knowledge transcends academic curiosity—it heralds a revolution in how we approach our collective well-being.

As we begin, think of the last time you stood before a tapestry. From a distance, it's a cohesive image, but as you draw nearer, the individual threads emerge, each one vital to the integrity of the whole. Blood types are the threads of our biological tapestry, and with careful study, their patterns emerge, revealing stories of human migration, evolution, and survival.

Did you know that the mere presence of certain antigens on the surface of your red blood cells could be the whisper of ancestral migrations from thousands of years ago? Or that the absence of a protein could speak to a communal resilience to a long-vanquished plague? It's not just about whether you're A, B, AB, or O—it's about what those letters represent in the grand chronicle of humanity.

But let's not get ahead of ourselves. This book will take you step-by-step through the science, from the basics of blood type genetics to the latest in epigenetic research that suggests our environment could be influencing blood type expression in ways we're only beginning to understand.

And yet, amidst this sea of information, we will not lose sight of the individual. For isn't it the individual that is most affected by the collective knowledge? Imagine a day when a quick blood test at your doctor's office doesn't just determine your blood type for transfusion compatibility but also guides your diet, your risk assessments for certain diseases, and even your mental health strategies.

With each chapter, I will weave in stories of real people—their struggles and triumphs—illuminating the profound impact that blood type research can have on personal health and global medical practices. You will meet individuals whose lives have been transformed by this knowledge, and experts who have dedicated their lives to untangling the mysteries of blood.

We'll explore the controversies, too—the ethical quandaries and the social implications of categorising people in new ways. But fear not, for this narrative is not about creating new divisions; it is about discovering deeper unities and celebrating the diversity of life.

So, dear reader, I invite you to journey with me. Cast aside your preconceived notions and prepare to question everything you thought you knew about blood. Each sentence, each fact, each revelation is a step closer to understanding the true significance of what flows within our veins.

The exploration of blood types is the next frontier, and together we will cross it, uncovering mysteries hidden within ourselves and forging a future where blood ties matter beyond the realm of race, where knowledge empowers us to transcend our visible differences and embrace the invisible ties that bind us all. Welcome to "The Next Frontier: Blood Type Research."

Blood Types in Medicine: Personalised Healthcare

In a world teeming with diversity, the silent whisper of blood coursing through our veins holds the potential to revolutionise medicine. It's a narrative that has been long overlooked, yet the implications of blood type knowledge for personalised healthcare are profound and far-reaching. What if the key to unlocking optimal health, to pre-empting disease, and to discovering treatments lay not in the visible spectrum of our skin but in the type of blood that fuels our very existence?

The issue at hand is clear: medical treatments have often adopted a one-size-fits-all approach, assuming that what works for one will work for all. These neglects the nuanced differences in our biological makeup, particularly our blood types, which research is increasingly showing to have significant bearings on our health.

Should this oversight persist, we stand before a cascade of potential consequences. Picture a world where vital medications are less effective for some due to their blood type, or where certain blood groups are more susceptible to illnesses yet remain blindly unaware. The cost of ignorance could be measured in lives lost or in the diminished quality of life for countless individuals.

Yet, the horizon is not devoid of hope. The solution lies in the heart of personalised medicine, where treatments and healthcare strategies are tailored to the individual, taking into account their unique blood type. This approach is not a mere theoretical concept; it is a practical, actionable strategy that begins with education and awareness.

The implementation of personalised healthcare based on blood types would commence with a comprehensive mapping of blood type distributions and their associated health patterns. Medical professionals must be trained to consider blood type as a significant factor in diagnoses and treatment plans. Moreover, public health initiatives must prioritise blood type testing and awareness, providing individuals with the knowledge of their own blood type and its implications.

Evidence of the efficacy of blood-type-based healthcare is mounting. Studies have shown that individuals with certain blood types may be at higher risk for specific diseases, while others may enjoy natural protection against them. By adopting personalised healthcare strategies, the prediction and prevention of these diseases can be vastly improved.

While this book advocates strongly for personalised medicine grounded in blood type knowledge, it is important to consider alternative solutions. Some argue for a broader genetic approach, looking at the entire genome to inform healthcare decisions. Others suggest that lifestyle and environmental factors should take precedence. These are not mutually exclusive from blood type considerations, but rather complementary pieces of a holistic health puzzle.

As we delve deeper into the intricacies of blood types, let's take a moment to ponder. Have you ever considered how your blood type might be silently influencing your health right now? Can you imagine a future where a simple blood test could unlock a personalised health regimen designed just for you?

The language of blood types is simple, yet its message is complex. It is a language that does not rely on the adjectives of race or the adverbs of ethnicity. Instead, it speaks in the nouns of antigens and the verbs of genetic expression. It is a language that, once understood, could redefine the cadence of healthcare.

There are stories waiting to be told, stories of individuals who have experienced firsthand the transformative power of blood-type-informed healthcare. Their journeys underscore the urgency of integrating this knowledge into our medical systems.

The future of medicine is a symphony of individual notes, each played in harmony with the others. It is a future where blood tie extend beyond the realm of race, where the barriers of healthcare are dismantled to reveal a personalised path to well-being for every individual.

This book is an invitation to embark on a journey of discovery, to envision a world where the silent markers in our blood inform the symphonies of our health. It is a call to action, to embrace the potential that lies within us all, and to forge a future where personalised healthcare is not just a possibility, but a reality.

Beyond Blood: The Quest for Universal Donor Solutions

In the silent channels of our veins, a revolution brews—one that is poised to redefine the landscape of medicine and challenge the very concept of compatibility in the transfusion of life's essence: blood. The pursuit of universal donor solutions has been a quest as old as transfusion medicine itself, and within these pages, we chronicle not only its history but its transformative journey into the modern era.

Blood, that rich and vital fluid, has been the subject of awe and scientific inquiry since time immemorial. From the first transfusions in the 17th century, where the mysteries of compatibility were yet unknown, to the ground-breaking discovery of blood types by Karl Landsteiner in 1901, the path to understanding blood's complexities has been fraught with both triumph and tragedy.

As Landsteiner unveiled the blood groups, a new era dawned—one where the compatibility of donor and recipient came sharply into focus. The realization that not all blood was alike spurred further research, leading to the identification of the Rh factor, a critical marker that further refined the safety of blood transfusions. With each stride in knowledge, the dream of a universal blood type—a panacea to the limitations of blood type matching—grew ever more vivid.

However, the quest for universal blood is not uniform across cultures and regions. In some parts of the world where certain blood types are more prevalent, the search for a universal donor solution takes on a different shape. In Japan, for example, where Type B is common, the implications for blood supply and transfusion practices contrast those of the Western hemisphere where Type O is more abundant.

The modern interpretation of universal blood solutions lies in cutting-edge research that seeks to strip away the barriers of blood type incompatibility altogether. Scientists are experimenting with enzymatic processes that can convert any donor blood into a type that mimics Type O-negative, the universal donor. This innovation could potentially make the need for type matching obsolete and revolutionize emergency medicine and surgery.

Yet, this journey is not without its challenges and controversies. The complexity of the human immune system means that altering blood antigens could have unforeseen consequences. Ethical considerations also emerge as we grapple with the cost and accessibility of these universal donor solutions. Will they be available to all, or only to a privileged few?

A turning point looms on the horizon, as the first clinical trials of these universal blood solutions begin to surface. The questions are direct and pressing: Will the promise of these innovations be realized, or will they succumb to the hurdles that lay before them?

With every word, this narrative has been woven to not only inform but to captivate. Imagine a world in which a paramedic, poised to save a life, need not pause to consider blood type compatibility. Envision a future where the blood flowing through the veins of any individual could be replenished without the fear of rejection.

The rhythm of progress in this field is neither steady nor predictable, but it pulses with the relentless determination of those who seek to change the paradigm of transfusion medicine. It is a rhythm marked by the beat of innovation, the tempo of research, and the measure of hope.

We find ourselves at the cusp of a new era, where the very blood that defines us may also be the key to transcending our limitations. The quest for universal donor solutions is more than a scientific endeavour; it is a testament to the indefatigable spirit of inquiry that has always propelled humanity forward.

This book is not merely a recounting of history; it is an ongoing story of aspiration and the human condition. And so, as we turn each page, we share in the collective breath of anticipation for what the next chapter of blood science will reveal, reminding us that within us all flows the potential for universal healing.

The Social Impact of Blood Type Awareness

In an age where the threads of society are interwoven with myriad hues of culture, ethnicity, and identity, we stand at the precipice of a new understanding—one that delves beneath the skin to the very lifeblood of humanity. Blood, a universal trait shared by all, yet varied in its typing, emerges as a potential catalyst for unity, transcending the divisive lines drawn by race. It is here, within the pulsing vessels of our global community, that we explore the potential for blood type awareness to reshape our social fabric.

Venture back to the classrooms of yore, where the rudiments of human biology were outlined in black and white. The concept of blood types was introduced as a mere footnote, a scientific curiosity with little perceived impact beyond the medical sphere. Fast forward to today, and we stand corrected. The primary challenge we face is not a lack of knowledge but the underutilisation of this information in fostering social cohesion.

The consequences of ignoring this aspect of our biological identity are manifold. At the benign end of the spectrum lies the missed opportunity for a deeper connection between individuals. On the darker side, it perpetuates the cycle of racial bias that afflicts our society, overshadowing our shared humanity with superficial divisions.

It is time to propose a fresh perspective: What if the awareness of blood types could become a bridge, a neutral ground upon which to build understanding and empathy? Could such a focus divert attention from racial differences and instead highlight our common biological bonds?

The solution begins with education. Imagine a curriculum enriched with the knowledge that while we may differ in appearance, the blood coursing through our veins speaks a universal language. Through this lens, we could teach our children that every human life is equally precious, each one capable of giving and receiving life through the gift of blood.

The implementation of this vision would require a concerted effort across multiple sectors. Educational systems would need to integrate blood type awareness into their programs. Media campaigns could promote stories of lives saved through blood donations that cross racial divides. Workplaces could host blood drives that emphasise the importance of all blood types, subtly reinforcing the message of unity.

Evidence of such a solution's efficacy can be gleaned from countries like Japan, where blood type is commonly believed to reveal personality traits. While the science behind these beliefs is debatable, the social impact is clear: it provides a platform for conversation and connection that transcends racial backgrounds.

But let us not be myopic, alternative solutions merit consideration. Some advocate for community-building activities that focus on shared interests or challenges, rather than biological traits. Others highlight the importance of traditional diversity and inclusion training. While these approaches are valid and valuable, they operate in the realm of the visible, the external. Blood type awareness digs deeper, tapping into the invisible threads that bind us.

Could it be that the very essence of life, the blood that flows within us, holds the key to dismantling the barriers that divide us? Can we dare to dream of a world where, in times of need, the question is not "Who are you?" but rather "What can I do?"

As the ink dries on these reflections, a singular truth resonates we are more alike than we are different. The heartbeats of seven billion people drum a collective rhythm, a reminder that beneath the surface, our blood matters beyond the realm of race. This is not a call to overlook the beauty of diversity but to cherish it with deeper understanding and respect.

The journey towards unity is not without its obstacles, but the path is illuminated with the light of knowledge and the hope of a more inclusive world. Let this be the legacy we pursue—one where our children look beyond the colour of skin to the content of character and the compatibility of blood.

We must encourage dialogue, ignite curiosity, and foster camaraderie based on the most fundamental level of human biology. The story of blood types is not just a tale of antigens and transfusions; it is a narrative of interconnectedness, a chronicle of humanity's shared journey through the ages.

As the pages of this book turn, each word is an invitation to consider a future where blood types, not racial stereotypes, are at the forefront of our social consciousness. It is a future that is not only possible but necessary for the evolution of society.

In the silent channels of our veins, a revolution brews—one that could not only redefine the landscape of medicine but also challenge the very fabric of social dynamics. It's time to let the silent channels speak and to listen with intent. For in the end, the blood that unites us is far more eloquent than the prejudices that divide.

A World United by Blood: A Vision for the Future

Imagine, if you will, a world where the commonality of our blood eclipses the divisions of our skin—a world in which the shared life force that pulses within us serves as the foundation for unity and understanding. This is not a mere fantasy; it is a future within our grasp, and through the pages of this book, you will embark on a transformative journey towards that horizon.

Dare to envision schools where children learn not only the ABCs but also the ABs and Os, where an understanding of blood types fosters a sense of common humanity. Picture a society where, in times of crisis, the rallying cry is not an us-versus-them but a life-affirming call to action that knows no boundaries of race or creed. This is the bold promise of "A World United by Blood: A Vision for the Future," a tome that will not only enlighten you but also inspire you to become an architect of this new, inclusive world.

As a medical science student, I have delved deeply into the wonders of human biology, and I stand before you with a message of hope grounded in the very essence of what makes us human. The methodologies I will unveil are not abstract theories but practical, actionable strategies that have the power to weave a tapestry of unity from the threads of our diverse blood types.

How can blood—a substance so often associated with injury and trauma—become a symbol of healing and connection? Let me guide you through the science and the stories that answer this question. From the intimate tales of lives saved through the selfless act of blood donation to the global narratives of communities united by the shared need for this life-giving resource, the proof of our interconnectedness is undeniable.

But you might wonder, how can such a vision overcome the deep-rooted scepticism and the entrenched racial divides that have long plagued our societies? It is a valid concern, and I confront it head-on by presenting not just ideals, but evidence. The facts are clear: while our blood types differ, the need for blood unites us all. Every time a person donates blood, they are potentially saving the lives of people from every race, every background, every walk of life. This is the irrefutable truth that forms the bedrock of our shared humanity.

Envision with me a world where blood type awareness is not a niche interest of the medically inclined but a common knowledge that promotes empathy and solidarity. Through this book, you will travel a path that leads from personal enlightenment to societal transformation, a journey that will challenge you to rethink the way we perceive our own identities and those of others.

Your commitment to reading these pages is not merely an investment in your own understanding but a step towards a collective awakening. The value of this knowledge transcends the personal; it has the power to change lives, to save lives, and to create a new paradigm of unity that our fractured world so desperately needs.

Imagine a day when the question "What's your type?" sparks a conversation about more than just compatibility—it becomes an opportunity to discover and celebrate our shared biology. Throughout this book, I will take you deep into the heart of what it means to be human, beyond the superficial markers that have divided us for far too long.

In this envisioned future, blood donation centres become places of communion, where diverse communities come together with a common purpose. Through vivid imagery, I will transport you to these centres of hope, where the rich tapestry of humanity is on full display, and the act of giving blood becomes a powerful statement of unity.

Can you picture it? A world where your blood type is a badge of honour, a symbol of your ability to reach across the divide and touch another's life in the most profound way possible. As you turn these pages, let the stories and the science wash over you, and allow yourself to be swept up in the possibility of a world united not by race, but by the very blood that gives us life.

And so, I ask you directly: Are you ready to be part of this vision? To embrace the simple, yet profound truth that our blood matters beyond the realm of race? To join the ranks of those who see beyond the colour of skin and recognize the universal language spoken by our veins?

By the end of this book, you will not only understand the power of blood type awareness; you will be an advocate for its potential to heal the wounds of division. The journey will be one of discovery, of challenge, and ultimately, of hope. It is a journey we will take together, with each chapter laying the groundwork for a more united humanity.

In the silent channels of our veins, a revolution brews—one that could not only redefine the landscape of medicine but also challenge the very fabric of social dynamics. It's time to let the silent channels speak and to listen with intent. For in the end, the blood that unites us is far more eloquent than the prejudices that divide.

CHAPTER

13

The Global Blood Dialogue

Imagine a world where the lines that divide us fade into the background, where humanity's collective heartbeat echoes louder than the discord of our differences. This is not just a distant dream—it's a tangible reality within our grasp. 'Blood Ties: Beyond Race' is not merely a book; it's a promise of a new dawn, an epoch where every drop of blood donated weaves the fabric of a united humanity.

As a medical science student, I've delved deep into the life-saving tapestry of blood donation. Through the chapters of this book, you will uncover the powerful methodologies and heart-stirring narratives that underscore the urgent need for a global initiative to augment blood donations.

You'll be privy to cutting-edge research, transformative case studies, and the unspoken bond that blood donation forges among strangers.

Perhaps you're questioning the impact of a single individual in this grand scheme or doubting the difference that one more book can make in the vast library of human knowledge. Let me dismantle those doubts. Each page of this book is imbued with the fervent belief that one person's resolve can send ripples through the ocean of humanity. I stand before you not just as an author, but as a testament to the life-altering power of action—one that is amplified when echoed by millions.

Envision with me a journey of transformation, where you're not just a passive reader but an active participant in a global crusade. Picture a future where blood shortages are a relic of the past, and every individual, regardless of race or creed, has access to the lifeblood they need. This is the transformation that 'Blood Ties: Beyond Race' promises to catalyse. This book is your compass, guiding you through the intricate maze of scientific innovation and human empathy.

By the time you turn the last page, you will not only have gained a wealth of knowledge, but you will also be armed with the tools to make a real difference. You'll carry the torch of this noble cause, lighting the way for others to follow. That is the commitment you make by joining this movement—a commitment to life, unity, and the unassailable bonds of blood ties.

And so, my fellow stewards of life, I call upon you to arm yourselves with compassion and courage as we embark on this vital mission. For the essence of our humanity is not found in the solitude of self, but in the collective pulse of society's heart as it beats in unison for a cause greater than any single life.

Now, let us delve into the heart of the matter, where facts meet passion and scepticism meets enlightenment.

Why is there such a dire need for blood donations, you may wonder? Every two seconds, someone in the world needs blood. From the victims of unforeseeable disasters to the silent warriors battling chronic diseases, the demand is relentless and unforgiving. Yet, the supply dwindles, hemmed in by misconceptions and unwarranted fears. Herein lies the crux of our endeavour—to bridge this chasm with knowledge, to replace fear with understanding, and to transform hesitation into action.

Consider the mother cradling her child with thalassemia, a genetic blood disorder requiring regular transfusions. Picture the trauma surgeon, her brow creased in focus, as she urgently calls for a unit of O-negative blood to save a life teetering on the edge. These are not just scenarios; they are daily realities etched into the lives of countless individuals. They are a clarion call for us, a summons to rise to the occasion and extend a lifeline.

Now, I ask you, dear reader, to ponder a question: What if that lifeline were to fray and snap? The consequences are not merely uncomfortable to think about—they are catastrophic to comprehend. The absence of adequate blood supply means the difference between life and death, between hope and despair. And yet, the solution is so simple, so profoundly within our reach. One hour of your time, one donation, can save up to three lives. Can there be a more compelling call to action?

In the following sections, you will encounter individuals whose lives have been irrevocably changed by the gift of blood donation. You will hear their voices, raw and unfiltered, as they express their gratitude for this priceless gift. You will learn of the scientific breakthroughs that have made blood donation safer and more efficient than ever before, and you will discover how cultural barriers can be dismantled, one donation at a time.

To the sceptics who believe that their contribution is but a drop in the ocean, remember that the ocean itself is made of drops. Each chapter of this book serves as a testament to the collective power of individual actions. Through vivid imagery and compelling narratives, you will see the faces of those you can help and feel the gratitude of lives saved.

You will not find jargon or incomprehensible medical lingo within these pages. Instead, you will find simple language that speaks directly to the heart, stirring the spirit and compelling action. You will find rhythm and cadence in the stories told, a harmony that reflects the pulse of life itself.

So, let us not delay. Let us heed the call to arms—for in our arms flows the blood that binds us, the blood that can heal wounds, save lives, and transcend the arbitrary boundaries of race. Together, let us forge a legacy of compassion, a testament to our shared humanity, and a world where blood ties extend beyond race, embracing the very essence of what it means to be human.

Welcome to 'Blood Ties: Beyond Race'. Welcome to the journey of a lifetime.

Blood Rights and Global Ethics

In a world where equity and access form the pillars of a just society, a silent inequity persists, often overlooked by the global discourse on human rights—the universal right to blood access. At the core of our shared humanity, beyond the lattice of politics and the mosaic of cultures, lies a fundamental need that knows no borders: the need for safe, available blood.

The gravity of this need cannot be overstated. Every minute, somewhere in the world, someone—be it a young woman in childbirth, an accident victim, or a surgery patient—faces the grim prospect of a dwindling blood supply. The harsh reality is that the scarcity of blood in medical facilities is not just a local issue; it is a global crisis that calls for an immediate and coordinated response.

Why, then, in an age of unprecedented medical advancement and global connectivity, does this crisis persist? The answer lies tangled in a web of ethical, cultural, and logistical challenges that span the breadth of our diverse global community. To untangle this web, we must confront hard truths and ask even harder questions.

Imagine, if you will, a child suffering from leukemia in a remote village, her lifeline tethered to the availability of compatible blood. Picture the furrowed brows of doctors in a bustling city hospital, as they grapple with the ethical dilemma of allocating scarce blood resources. These scenarios are far from rare; they are the daily reality for many, a reality that is as complex as it is heartbreaking.

What if, in that remote village, there was no blood available? What if, in that city hospital, the choice of who receives blood and who does not is dictated not by need, but by social status, financial means, or worse, discrimination? The stakes are indeed high, for the cost of inaction is measured in human lives.

This book seeks not just to highlight the problem but to illuminate the path to solutions. From the intricate policies governing blood banks to the grassroots movements striving for change, we will explore the multifaceted approaches that can turn the tide of this crisis. We will delve into the ethical debates that surround blood donation and distribution, and we will examine the role of international law and global governance in ensuring that the right to blood access is upheld.

To the weary, the cynically inclined, who might ask, "What can be done in the face of such a vast and complex issue?"— this book answers with a resounding "Much." The topics ahead will chart the course of action, from individual commitment to systemic overhaul, and from community education to international collaboration.

Through the power of storytelling, we will traverse continents and cultures to meet the unsung heroes who are making a difference. From the donors who roll up their sleeves to give a part of themselves to strangers, to the activists fighting to dismantle the barriers that prevent marginalised communities from donating blood, their stories will inspire and challenge us.

These narratives will be interwoven with expert insights, shedding light on the scientific innovations that promise to revolutionise blood donation and storage. We will explore ground-breaking research that aims to create a future where synthetic blood could alleviate shortages, and where the very concept of 'blood type' may become obsolete.

We will confront the uncomfortable questions: How do we balance the right to privacy of donors with the need for safe blood? What are the ethical implications of compensating donors in regions where voluntary donations are not meeting demand? How can we ensure that the global policies on blood access are fair and just, not only in letter but in spirit?

The book will take you on a journey through the evolving landscape of global blood ethics, where the moral compass must be recalibrated to account for the vast inequalities that exist. The solutions proposed herein will not be mere platitudes, but actionable strategies rooted in compassion, equity, and scientific rigor.

And so, with unwavering resolve, we must engage with the message of this book. We must allow it to stir our consciences, to move us from complacency to action. For within its pages lies a blueprint for a world where no one is left to bleed in the absence of that most vital substance—blood.

As we turn each page together, let us commit to the vision it presents—a world where blood ties do not bind us to a race or nation, but to a single, global family. Let us rise to the challenge, armed with knowledge and united by the common pulse that courses through our veins. For it is only through our collective will that we can ensure blood rights are recognised not as a privilege for some, but as a global ethical imperative for all.

Welcome to the next chapter in our journey. Welcome to 'Blood Rights and Global Ethics'.

International Blood Policies: A Comparative Study

In the shadowed corners of the world's healthcare systems, where policy meets practice, lies the intricate and often opaque realm of international blood policies. With each nation sculpting its framework based on a complex blend of culture, ethics, and necessity, understanding these policies becomes a task of unwinding the threads of global medical tapestry. This section of our exploration dives deep into the fabric of international blood policies, comparing and contrasting them to glean lessons that could inform a path toward a universal standard of blood equity and ethics.

The significance of such a study stretches far beyond academic interest. Blood, the elixir of life, courses through the veins of every human being, and the policies that govern its donation, storage, and distribution are as vital as the heartbeat itself. Here, we seek to unravel the rationale behind the policies of disparate nations and the insights they offer into a collective approach to blood management that honours the sanctity of life above all else.

Why compare blood policies across borders? The answer lies in the quest for universal best practices that transcend cultural and political divides. By examining the successes and failures of various systems, we can distil the essence of what works and what doesn't, forging a path to a future where blood scarcity is a relic of the past.

The criteria for our comparison are meticulously chosen to cover the spectrum of policy facets: from donor eligibility and recruitment strategies to the ethical considerations of donor compensation; from the technicalities of blood screening and processing to the logistics of distribution and accessibility.

When we juxtapose the policies of, say, the United States with those of Japan, we find intriguing similarities. Both nations rely heavily on voluntary, unpaid blood donations as the cornerstone of their blood supply. This shared belief in altruism underscores the acknowledgment that the act of giving blood is, at its core, a humanitarian one.

Yet, a closer look reveals stark contrasts in execution. The United States of America operates under a decentralised system, with a myriad of organisations collecting and distributing blood, whereas Japan benefits from a more centralised approach, with the Japanese Red Cross Society (JRCS) playing a pivotal role. How do these structural differences impact efficiency and equity in blood access? The answer is not a simple one, but it lies in the stories of lives touched by each system.

Our analysis delves into the heart of these policies, unravelling the threads of tradition and modernity that weave together to form the safety nets of societies. We see how, in countries like Brazil, innovative blood donor recruitment campaigns harness the power of social media to reach potential donors, especially among the youth. The broader implications of such initiatives are profound, demonstrating the potential for technology to bridge gaps in blood supply.

But what of the contemporary relevance of these policies? How do they stand up to the challenges of a world in turmoil—a world where pandemics and natural disasters test the resilience of blood supply systems? We find that in some cases, like in the aftermath of the 2010 earthquake in Haiti, international aid and swift policy adaptations played a critical role in meeting the sudden surge in demand for blood transfusions.

The narrative then shifts to a sombre tone, as we confront the darker side of blood policies. Instances where discrimination—based on race, or socioeconomic status—creeps into the eligibility criteria, painting a picture of exclusion that contradicts the spirit of universal blood rights.

Yet, through the gloom, rays of hope emerge. We recount the poignant tales of individuals whose lives were saved by the very policies they once thought indifferent to their plight. These stories are not merely recounted; they are shown through the eyes of the survivors, the tireless workers, and the policymakers themselves.

In a world fraught with inequities, the lessons gleaned from this comparative study are a clarion call to action. We are compelled to ask ourselves: How can we shape a global blood policy that honours the dignity of all individuals? What can we learn from the successes and pitfalls of existing systems?

This quest for answers is not a solitary journey. It is a collective endeavour, requiring the input and collaboration of all stakeholders—policymakers, healthcare professionals, donors, and recipients alike. As we forge ahead, let us take these insights and use them to craft a world where the right to blood is as inalienable as the right to life itself. The pages that follow will not only document this quest but serve as a testament to our shared humanity—a humanity that knows no borders when it comes to saving lives.

Blood as a Unifier

Embark upon a profound exploration with me, and together we shall uncover the undeniable truth that blood is the ultimate emblem of our shared humanity, an unbreakable bond that eclipses the superficial barriers of race. This book promises to illuminate your mind and stir your heart, revealing the powerful connections that link us all.

From the bustling streets of metropolitan cities to the secluded paths of remote villages, the rich crimson that flows in our veins is the same. This universal life force carries the potential to unite us in ways that we are only beginning to understand, and it is my mission to reveal the intricate ways in which our blood can serve as a bridge rather than a barrier.

Through compelling narratives, scientific insights, and thought-provoking discussions, I will guide you through the complexities of our biological commonalities. We will explore the marvels of genetics, the intricacies of blood types, and the miraculous ways in which blood not only sustains life but also has the power to heal and connect us.

Yet, you may wonder, how can such a claim stand in a world where racial divides appear insurmountable? I address this scepticism head-on by presenting evidence of our interconnectedness, dismantling the myths that have long fuelled division. I will show you the profound impact of recognising our shared bloodline, how it fosters empathy and dismantles prejudices.

Picture a world where the lines that separate us fade away, leaving a clear view of the common ground we all share. This vision is not a distant utopia—it's a reality that can be achieved through the recognition and celebration of our shared blood ties. By recognising the strength in our biological bonds, we can lay the foundation for a more compassionate and inclusive society.

This journey will not only broaden your perspective but will also challenge you to reconsider the way you see yourself and others. It is a call to action, urging you to become an advocate for a world where race is acknowledged as just one aspect of our identity, not the defining factor.

As you turn the pages of this book, you will feel a growing sense of connection with your fellow humans. This realisation is not merely enlightening; it is transformative. It has the potential to change the way we interact with one another, the way we formulate policies, and the way we envision our future as a species.

The commitment you make by engaging with this content is significant. You are not just reading a book; you are partaking in a movement towards unity and understanding. The life-changing potential of embracing our blood ties extends far beyond the individual—it has the power to alter the course of societies and, ultimately, the fabric of humanity.

Now, let us delve deeper into this exploration.

How often do we consider the power of the blood coursing through our veins? It's the universal currency of life, an element so fundamental that its significance is often overlooked in the noise of daily existence. Yet, within this quiet stream flows the potential to challenge and change our deepest convictions about identity and belonging.

With a background in medical science, I have been uniquely positioned to study the marvels of human blood. My education at Edith Cowan University has not only equipped me with the knowledge of the biological aspects of blood but also impressed upon me the profound ethical implications of how we view and utilise this precious resource.

Through the lenses of science and ethics, we will examine the remarkable stories of blood donation and transfusion, stories that transcend boundaries and forge unspoken bonds between strangers. Have you ever pondered the invisible threads that connect a young woman in Sweden with a man in the heart of Africa through the simple act of a blood transfusion? This book brings such miraculous connections to light, showcasing the undeniable evidence of our shared humanity.

Moreover, we will look beyond the act of sharing blood to the deeper symbolism it carries. What does it mean to give a part of oneself to another? How does this act challenge the divisive narratives of race and ethnicity that pervade our societies?

The language of blood is simple, yet it speaks volumes. It does not whisper of differences; it shouts of unity. It does not discriminate by skin colour; it discriminates only by compatibility, which knows no racial boundaries. In this book, I present the biological facts, the emotive stories, and the philosophical musings that collectively argue for a new consciousness—one that sees blood as the tie that inherently binds us beyond any constructed racial divides.

In our exploration, we will also encounter the heroes who work tirelessly within the intricate networks of blood banks and hospitals, often unheralded stewards of life. Their dedication and innovation in the face of adversity serve as a testament to what can be achieved when we prioritise our common well-being over our perceived differences.

Questions will arise, challenging us to think and rethink our preconceptions. What if our first instinct in seeing another person was to recognise them as a potential blood brother or sister, rather than as a member of a different race? How would our world change if we acted on the principle that, in times of need, we all bleed the same?

My dear reader, as you immerse yourself in the pages to come, let the vivid imagery of our shared bloodstream enliven your imagination and broaden your worldview. This is not a narrative of division; it is a celebration of connection.

And so, let us begin this journey with open hearts and minds, ready to embrace a vision of humanity united not by the colour of our skin but by the blood that gives us life. Welcome to 'Blood Ties: Beyond Race', where we discover that what unites us is far more powerful than what divides us. Welcome to the profound realisation that, indeed, blood is the great unifier.

The Blood Ties Movement: A Vision for Change

In the pages that have preceded this moment, we have journeyed together through the rich tapestry of our shared humanness, carried within the life-giving rivers that flow within us all. The red rivers of life unite us in silent kinship, speaking a language of commonality that outshouts the cacophonous narratives of division. It is now time to chart the course forward, to lay out the blueprint for a movement that embodies the principles of blood-based solidarity. This is the vision for change—the genesis of the Blood Ties Movement.

The goal is unequivocal and bold: to cultivate a global community that recognises and celebrates the primordial connection of blood ties over superficial and divisive constructs of race and ethnicity. This movement seeks to transform the societal fabric by fostering a profound and pervasive recognition of our biological oneness.

Before embarking on this transformative odyssey, one must gather the necessary materials or prerequisites. They are not tangible items but rather intangible yet crucial elements: an open mind, a compassionate heart, the willingness to engage in difficult conversations, and the courage to challenge entrenched beliefs.

Allow me to provide a broad overview of the steps we will undertake. We will begin by nurturing awareness and understanding, then move on to building community and solidarity.

Following this, we will engage in activism and advocacy, and finally, we will work towards systemic and policy reforms.

Let us now dive deeper into each phase, examining the detailed steps that will guide us to our destination:

Awareness and Understanding:

First, we must engage in educational efforts that illuminate the scientific truths of human blood and its universality. Host seminars, create online content, and encourage dialogue in schools and communities. Share the emotive stories of blood donation and transfusion that transcend borders, and let these narratives serve as a conduit for empathy.

Community and Solidarity:

Next, we must connect individuals through shared experiences and initiatives. Organise blood donation drives that double as community-building events. Foster networks of support that not only provide aid but also promote the recognition of our mutual bonds.

Activism and Advocacy:

As our community grows, we must lend our voice to the voiceless and advocate for policy changes that reflect our values. Collaborate with organisations that work towards racial equality and health equity. Use the power of social media to spread our message and galvanise action.

Systemic and Policy Reforms:

Finally, we must strive to influence systemic changes. Work with lawmakers and institutions to ensure that the principles of blood ties are reflected in policies and practices. Push for the implementation of educational curriculums that promote an understanding of our shared biology.

On this voyage, I offer you these tips and warnings: Stay patient, for change is often slow. Be resilient in the face of resistance. Remember, the strength of our conviction must match the magnitude of the change we seek.

To test or validate the successful completion of each step, look for measurable outcomes. Are awareness and understanding increasing? Is community participation growing? Are advocacy efforts resulting in tangible policy discussions?

Should you encounter challenges, as you inevitably will, do not be disheartened. Troubleshoot by revisiting your strategies, seeking feedback, and being adaptable. Remember that every movement faces opposition, but it is the perseverance and adaptability of its members that ultimately leads to progress.

In the spirit of varying sentence openers, let us pose a direct question: What would the world look like if we all saw each other as part of an extended family, linked by the blood that courses through our veins?

Imagine, if you will, the image of a child offering a helping hand to another, not because they share the same skin colour, but because they recognise in each other the same crimson life force. Use strong verbs and nouns to convey this powerful scene, for it is in the simplicity of such acts that the most profound changes are born.

Occasionally, let us pause and reflect in a one-sentence paragraph: Unity is our birthright, written in the indelible ink of our blood.

Incorporate quotations to add authenticity to our cause. As the great poet Maya Angelou once said, "We are more alike, my friends, than we are unalike." Let this simple yet profound truth resonate within the movement.

And always, show, don't tell. Do not merely state that blood ties are strong; show their strength through the stories of those whose lives have been saved by the blood of strangers, of communities that have come together to celebrate this shared bond.

In conclusion, the Blood Ties Movement is not a fleeting campaign; it is a perpetual testament to our unity. It is a call to look beyond the seen to the unseen, to recognise that beyond race, we are bound by something far more enduring—the very essence that gives us life. Join us in this vision for change, where blood ties are not merely acknowledged but revered as the foundation for a more unified and compassionate world.

The Power Within Us!

(song)

In the depths of our veins, a secret lies untold,
A power within us, a story yet to unfold,
It's the blood that flows, through our every cell,
A force so mighty, that no one can quell.

Oh, the power within, our blood group so rare,
A gift from our ancestors, beyond compare,
From A to B, O to AB, we stand tall,
United by the power, that runs through us all.

Type A, the warriors, with hearts full of fire,
Driven by passion, they never tire,
Their blood runs with purpose, compassion their guide,
A force to be reckoned with, side by side.

Type B, the adventurers, with spirits so free,
They chase their dreams, wherever they may be,
Their blood carries courage, a thirst for the unknown,
They conquer mountains, with strength they've shown.

Oh, the power within, our blood group so rare,
A gift from our ancestors, beyond compare,
From A to B, O to AB, we stand tall,
United by the power, that runs through us all.

O, the universal donor, selfless and kind,
Their blood saves lives, a treasure we find,
With every donation, they bring hope anew,
A superhero's power, shining through.

And AB, the enigma, a blend of two worlds,
With empathy and wisdom, their flag unfurled,
Their blood holds the secrets, of both A and B,
A bridge between differences, for all to see.

Oh, the power within, our blood group so rare,
A gift from our ancestors, beyond compare,
From A to B, O to AB, we stand tall,
United by the power, that runs through us all.

So let us celebrate, this power we possess,
The strength in our blood, that we must confess,
For united we stand, in this diverse crowd,
The power within blood groups, forever proud.

Glossary

Agglutination: The clumping together of particles, such as red blood cells, in response to specific antibodies, creates a cohesive mass.

Alleles: Variations in inherited traits are determined by different forms of a gene known as alleles.

Amalgamation: The process of merging and combining different elements into a unified whole result in amalgamation.

Amniocentesis: A prenatal diagnostic test that involves extracting a sample of amniotic fluid to identify genetic disorders.

Anecdotal: Instead of relying on scientific evidence, anecdotal information is based on personal accounts or stories.

Antagonists: Substances that counteract or inhibit the effects of another substance are known as antagonists.

Antibody: The immune system produces proteins called antibodies to neutralise or destroy foreign substances.

Anticoagulants: Substances that prevent blood from clotting are referred to as anticoagulants.

Anthropologists: Scientists who study the development, societies, and cultures of human beings are known as anthropologists.

Carbon dioxide: The lungs exhale carbon dioxide, which is a waste product of cellular respiration.

Carbohydrates: Carbohydrates are one of the three macronutrients (along with proteins and fats) and are a primary source of energy for the body.

Cardiac: Relating to the heart, anything pertaining to the heart is considered cardiac.

Cardiovascular: Anything that relates to both the heart and blood vessels is considered cardiovascular.

Catheter: A thin tube that is inserted into a blood vessel or body cavity to administer medications, draw blood, or perform medical procedures.

Cell: The basic unit of life, capable of carrying out various functions within the body.

Cellular: Relating to cells or the basic structural and functional units of living organisms.

Chronological: Arranged in the order of time, events or information that follows a chronological order.

Circulatory system: Responsible for transporting blood, oxygen, and nutrients throughout the body, the circulatory system plays a vital role.

Coagulation: The process of blood clotting to prevent excessive bleeding is known as coagulation.

Dichotomy: A division or contrast between two opposing or entirely different things is referred to as a dichotomy.

Digestive: Relating to the process of breaking down food and absorbing nutrients in the body.

Epidemic: The rapid spread of a disease to a large number of people within a given population.

Epidemiological: Refers to the study of the distribution and determinants of health-related events, conditions, and diseases in populations.

Eugenics: A movement that aimed to improve the genetic quality of the human population through selective breeding and sterilisation.

Eucharist: (holy communion or the Lord's Supper): A Christian sacrament that commemorates the Last Supper of Jesus Christ.

Gastronomic: Relating to the art or science of good eating and cooking.

Genesis: The origin or beginning of something.

Genetic drift: Random changes in the frequency of gene variants in a population over time.

Genotype: The genetic makeup of an individual that determines their blood types.

Genome: The genome is the complete set of genetic material (DNA or RNA) of an organism. It contains all the information necessary for the development, functioning, and reproduction of the organism.

Haematology: The study of blood and blood disorders.

Haemolytic: Relating to the destruction of red blood cells.

Haemorrhaging: The escape of blood from a ruptured blood vessel.

Haemophilia: A genetic disorder characterised by the inability of the blood to clot properly. People with haemophilia experience prolonged bleeding and are at a higher risk of excessive bleeding from injuries or surgeries.

Hemochromatosis: A hereditary condition in which the body absorbs too much iron from the diet. The excess iron is stored in various organs, including the liver, heart, and pancreas, which can lead to organ damage.

Homogeneous: Consisting of the same or similar components throughout.

Hydrops fetalis: A condition in which excessive fluid accumulates in the foetus, causing severe swelling.

Immunological: Pertaining to the immune system and its response to foreign substances.

Immune system: The body's defence mechanism against infections and diseases.

Incompatibility: The inability of blood types to mix without adverse reactions.

Inflammation: The body's response to injury or infection, characterised by redness, swelling, and pain.

Intrauterine: Inside the uterus or womb.

Intravenous: Administering fluids or medications directly into a vein.

Jaundice: A yellowing of the skin and eyes due to liver dysfunction or excessive breakdown of red blood cells.

Juxtapose: To place two or more things side by side for comparison or contrast.

Ketogenic: Ketogenic refers to a diet or metabolic state characterised by the production of ketones in the body. Ketones are produced when the body breaks down fat for energy in the absence of sufficient carbohydrates.

Kaleidoscope: A tube-shaped optical instrument that produces changing patterns when rotated.

Labyrinthine passage: A complex and intricate pathway or route.

Lectins: Proteins found in many foods that can bind to carbohydrates and have various biological effects.

Leukaemia: A type of cancer that affects the blood and bone marrow. It causes the production of abnormal white blood cells, which crowd out healthy blood cells and impair the body's ability to fight infections.

Lexicon: The vocabulary or set of terms used in a particular language or subject.

Metabolic: Relating to the chemical processes that occur within living organisms to maintain life.

Melanocytes: Melanocytes are specialised cells found in the skin, hair, and eyes that produce the pigment melanin. Melanin is responsible for the colour of our skin, hair, and eyes.

Molecules: The smallest unit of a chemical compound, composed of atoms held together by chemical bonds.

Nutrients: Substances that provide nourishment and are essential for growth and maintenance of the body.

Obstetric: Relating to pregnancy, childbirth, and the postpartum period.

Oxygen: A gas essential for respiration and energy production in cells.

Organ: A specialised structure in the body that performs specific functions.

Organism: Any living entity, such as plants, animals, or microorganisms.

Phenotype: The observable physical or biochemical characteristics of an organism.

Phlebotomy: The process of drawing blood from a vein for diagnostic testing or therapeutic purposes. Phlebotomy is commonly performed by healthcare professionals, such as nurses or phlebotomists.

Phrenology: A pseudoscience that claimed to discern intellectual and moral traits from the shape of a skull.

Pheromones: Pheromones are chemical substances produced by animals and humans from their skin glands that influence the behaviour of attracting mates.

Pathogen: A pathogen is a microorganism (such as a bacterium, virus, fungus, or parasite) that can cause disease in a animals and humans.

Platelets: Small, disc-shaped cells in the blood that help with blood clotting. Platelets play a crucial role in stopping bleeding by forming clots at the site of an injury.

Prenatal: Relating to the period before birth.

Plasma: The liquid component of blood that carries cells, nutrients, hormones, and waste products throughout the body.

Red blood cell: Blood cells that carry oxygen to the body's tissues.

Saline solution: A solution of salt in water, often used for medical purposes.

Serum: The clear, yellowish fluid that remains after blood has clotted, containing antibodies and other proteins.

Sickle cell anaemia: A genetic disorder characterised by abnormal, crescent-shaped red blood cells. These cells can get stuck in blood vessels, leading to reduced oxygen flow and causing pain, organ damage, and other complications.

Sodium Citrate: A compound used as an anticoagulant in blood transfusions.

Thalassemia: A group of inherited blood disorders that affect the production of haemoglobin, the protein in red blood cells that carries oxygen. People with thalassemia have fewer healthy red blood cells, leading to anaemia and other symptoms.

Taxonomy: The science of classification, especially of organisms.

Terminal tuberculosis: The advanced stage of tuberculosis, a contagious bacterial infection that primarily affects the lungs.

Tissue: A group of cells that work together to perform a specific function in the body.

Traits: Characteristics or qualities that are inherited or acquired by an organism.

Transfusion: The process of transferring blood or blood products from one person to another.

Typology: The classification or categorisation of things based on shared characteristics.

Universal donor: A person with blood type O-negative, which can be transfused to individuals of any blood type.

Universal recipient: A person with blood type AB-positive, which can receive transfusions from individuals of any blood type.

Vein: Blood vessels that carry deoxygenated blood back to the heart.

White blood cell: Blood cells that play a crucial role in the immune system's defence against infections.

West Nile Virus: Refers to a mosquito-borne virus that can cause illness in humans and animals. It is primarily transmitted through the bite of infected mosquitoes.

Acronyms

ABO - ABO blood group system

AIDS - Acquired Immuno Deficiency Syndrome

CRISPR-Cas9 - Clustered Regularly Interspaced Short Palindromic Repeats

COVID-19 - Coronavirus Disease 2019

DNA - Deoxyribonucleic Acid

ECG - Electrocardiogram

HDN - Haemolytic Disease of the New-born

HIV - Human Immunodeficiency Virus

JRCS - Japanese Red Cross Society

NIV - The New International Version

RBC - Red Blood Cell

RH - Rhesus Factor

RNA - Ribonucleic Acid

Bibliography

Addi, G. J. (1959). Blood groups in Acute Rheumatism. Scottish Med. J. P 547.

Banks. (2020). Technical Manual (20th ed.). AABB

Brues, A. M. (1929). Tests of Blood Group Selection. Amer. J Forensic Medicine.

Cohn, C. S., Delaney, M., Johnson, S. T., Katz, L. M., & American Association of Blood.

Carey, N. (2020). Hacking The Code of Life: how gene editing will rewrite our futures. Icon Books Ltd.

Dissected Lives. (2020). Carolus Linnaeus: The Life and Works of the Father of Modern Taxonomy | Naming the World Grade 5 | Children's Biographies. Speedy Publishing LLC.

D'Adamo, P. J., & Whitney, C. (2000). Live Right 4 Your Type. Penguin.

D'Adamo, P. (1982). Blood types and diseases, a review. Clinical Rounds Presentation, Bastyr University.

D'adamo, P., & Whitney, C. (2016). Eat right 4 your type. Century.

D'Adamo, P. J. (1996). Eat Right for Your Type: The Individualized Diet Solution to Staying Healthy, Living Longer & Achieving Your Ideal Weight. Penguin Putnam.

Daniels, G. (2013). Human blood groups. John Wiley & Sons.

Felicia Drury Kliment. (2006). Eat right for your metabolism: how the right foods for your type can help you lose weight. McGraw-Hill.

Freed, S. (2013). Power in the Blood. Baker Books.

Giangrande, P. L. F. (2000). The history of blood transfusion. British Journal of Haematology, 110(4), 758–767. https://doi.org/10.1046/j.1365-2141.2000.02139.x

Giessen, C. (2001). Diets and why most don't work: personalizing your diet to suit your blood type and your body needs. New Century Press.

Grave errors: Spooky cures and creepy medical missteps from the past. (n.d.). AAMC. Retrieved January 23, 2024, from https://www.aamc.org/news/grave-errors-.spooky-cures-and-creepy-medical-missteps-past

Ito M. (2002). The diagnosis from the pathological viewpoint of a blood disease. International journal of hematology, 76 Suppl 2, 2–5. https://doi.org/10.1007/BF03165076

Irus Braverman. (2018). Gene editing, law, and the environment: life beyond the human. Routledge, Taylor & Francis Group.

LaValle, J. B. (2013). Your Blood Never Lies: How to Read a Blood Test for a Longer, Healthier Life. In Google Books. Square One Publishers. https://books.google.com.au/books/about/Your_Blood_Never_Lies.

Latham, M. A. (2017). Sherlock Holmes - A Betrayal in Blood. Titan Books (US, CA).

Lee, T. F. (2013). The Human Genome Project. Springer.

Marcus, D.M. (1972). The ABO and Lewis Blood-Group System. New York: Bantam.

Mujahid, A., & Dickert, F. (2015). Blood Group Typing: From Classical Strategies to the Application of Synthetic Antibodies Generated by Molecular Imprinting. Sensors, 16(1), 51. https://doi.org/10.3390/s16010051

Nawata K. (2014). Shinrigaku kenkyu: The Japanese journal of psychology, 85(2), 148–156. https://doi.org/10.4992/jjpsy.85.13016.

O'Shea, T. (2009). The Sanctity of Human Blood.

Red Cross Australia. Blood saving lives. https://www.redcross.org.au/sites/annual-reports/2014/year-in-review/blood-saving-lives. Date accessed, January 23, 2024.

Shadrack Mvunabandi. (2013). The Communicative Power of Blood Sacrifices.

Thomas, T. Gaillakd. (1878). Intra Venous Injection of Milk as A Substitute Foe the Transfusion Of Blood. Illustrated By Seven Operations. National Institutes of Health. New York. may, 1878. Retrieved from: https://digirepo.nlm.nih.gov/ext/dw/101670861/PDF/101670861.pdf.

Toshitaka Nomi, & Besher, A. (1988). You are your blood type: the biochemical key to unlocking the secrets of your personality. Pocketbooks.

Toshitaka Nomi. (2017). You are Your Blood Type. Noura Books

The Nobel Prize. (n.d.). The Nobel Prize in Physiology or Medicine 1930. NobelPrize.org. https://www.nobelprize.org/prizes/

Vasey, C. (2015). The Spiritual Mysteries of Blood. Simon and Schuster.

Watson, W. (2011). Personality & Blood Type.

Winslow, R. M., Vandegriff, K. D., & Intaglietta, M. (2012). Blood Substitutes. Springer Science & Business Media.

Walker, K. (1958). The Story of Blood.

Wijangco, D. A. (2017). Sanctity of Blood.

Year in review - Australian Red Cross. (2013). Redcross.org.au. https://www.redcross.org.au/sites/annual-reports/2014/year-in-review/blood-saving-lives/

Author's Biography

Kenyatta National Hospital: Accident & Emergency Wing

Abdisa's journey began in the humble town of Bojji Dirmaji, nestled in the heart of Western Wellega, Ethiopia. Growing up in this tight-knit community, he was exposed to a world of storytelling that ignited his imagination and shaped his unique perspective on life.

But Abdisa's ambitions extended beyond the realm of storytelling. With a deep-rooted desire to make a difference in the world, he pursued a career in healthcare. Graduated with great distinction as a Clinical Nurse in Addis Ababa, he demonstrated his unwavering dedication to the well-being of others.

Yet, Abdisa's thirst for knowledge and his insatiable curiosity led him to embark on a new adventure. He set his sights on Edith Cowan University in Australia, where he is currently studying Medical Science. This dual role as both an aspiring author and a diligent student in the medical field epitomises his diverse interests and unwavering commitment to creating a positive impact on the world.

Abdisa's journey from a small town in Ethiopia to the halls of academia in Australia is a testament to his resilience, determination, and boundless potential. With his pen and stethoscope in hand, he is poised to leave an indelible mark on both the literary and medical world.

www.ingramcontent.com/pod-product-compliance
Lightning Source LLC
LaVergne TN
LVHW050910210926
848482LV00012B/391